In this book the authors illustrate the basic physics and materials science of conjugated polymers and their interfaces, particularly, but not exclusively, as they are applied to polymer-based light emitting diodes. The approach is to describe the basic physical and associated chemical principles that apply to these materials, which in many instances are different from those that apply to the inorganic counterparts.

The main aim of the authors is to highlight specific issues and properties of polymer surfaces and interfaces that are relevant in the context of the emerging field of polymer-based electronics in general, and polymer-based light emitting diodes in particular. Both theoretical and experimental methods used in the study of these systems are discussed. Generally speaking the discussions are concise and to the point, but the approach is such that any scientist entering this field will be able to use this book to become familiar with the most applicable theoretical and experimental methods employed, and to read and understand published papers in this field.

This book will be of interest to graduate students and research workers in departments of physics, chemistry, electrical engineering and materials science studying polymer surfaces and interfaces and their application in polymer-based electronics.

Conjugated polymer surfaces and interfaces

Conjugated polymer surfaces and interfaces

Electronic and chemical structure of interfaces
for polymer light emitting devices

W. R. Salaneck

S. Stafström

Department of Physics, IFM
Linköping University, Sweden

and

J.-L. Brédas

Service de Chimie des Matériaux Noveaux
Centre de Recherches en Electronique et Photonique Moléculaires
Université de Mons-Hainault, Belgium

CAMBRIDGE
UNIVERSITY PRESS

PUBLISHED BY THE PRESS SYNDICATE OF THE UNIVERSITY OF CAMBRIDGE
The Pitt Building, Trumpington Street, Cambridge, United Kingdom

CAMBRIDGE UNIVERSITY PRESS
The Edinburgh Building, Cambridge CB2 2RU, UK
40 West 20th Street, New York NY 10011–4211, USA
477 Williamstown Road, Port Melbourne, VIC 3207, Australia
Ruiz de Alarcón 13, 28014 Madrid, Spain
Dock House, The Waterfront, Cape Town 8001, South Africa

http://www.cambridge.org

First published 1996
First paperback edition 2003

A catalogue record for this book is available from the British Library

Library of Congress cataloguing in publication data

Salaneck, W. R.
Conjugated polymer surfaces and interfaces: electronic and chemical structure of interfaces for
polymer light emitting devices / W. R. Salaneck, S. Stafström and J-L. Brédas.
        p.   cm.
  Includes bibliographical references and index.
  ISBN 0 521 47206 7 (hardback)
  1. Light emitting diodes – Materials.   2. Polymers – Surfaces.   3. Surface chemistry.
4. Polymers – Electric properties. I. Stafström, S.   II. Brédas, J-L. (Jean-Luc), 1954–   .
III. Title.
TK7871.89.L53S35   1996
621.3815′2—dc20   95–38726 CIP

ISBN 0 521 47206 7 hardback
ISBN 0 521 54410 6 paperback

Contents

Preface

The recent flourish in activities, both in academia and industry, in research on polymer-based light emitting devices has prompted a series of studies of the surfaces of conjugated polymers, and the early stages of interface formation when metals are vapour-deposited on these conjugated polymer surfaces. The summary of works presented here is, to a great extent, the result of a long and comprehensive co-operation between both the Laboratory of Surface Physics and Chemistry, and the Molecular Theory group within the Laboratory for Theoretical Physics, at the Department of Physics (IFM) at Linköping University, Sweden, and the Laboratory for Chemistry of Novel Materials and the Center for Research on Molecular Electronics and Photonics, at the University of Mons-Hainaut, Belgium. The results brought forth within this collaboration represent the product of a comprehensive combined experimental–theoretical approach to the study of the electronic and chemical structure of conjugated polymer surfaces and interfaces. The scope of the output of such a combined approach is greater than the sum of the (theory and experimental) parts. Following about 10 years' work, we were asked to compile a summary of the highlights of our studies in this area. This monograph represents a response to that request.

For generation of this book, we acknowledge specific support from the Commission of the European Union within the Network of

Excellence on Organic Materials for Electronics NEOME). For participation in these activities, we acknowledge Commission support through the SCIENCE program (Project 0661 POLYSURF), the Brite/EuRam program (Project 7762 PolyLED), and the ESPRIT program (Project 8013 LEDFOS). Individually, our research activities have been supported by grants from the Swedish Natural Sciences Research Council (NFR), the Swedish National Board for Industrial and Technical Development (NuTek), the Swedish Research Council for Engineering Sciences (TFR), and the Neste Corporation, Finland; as well as by the Belgian Prime Minister's Office for Science Policy (SSTC) Programs 'Pôle d'Attraction Interuniversitaire en Chimie Supramoléculaire et Catalyse' and 'Programme d'Impulsion en Technologie de l'Information (contract SC/IT/22)', by the Belgian National Fund for Scientific Research FNRS, and by an IBM Academic Joint Study.

Chapter 1

Introduction

Although electroluminesence from organic materials[1] has been
known for a long time, research on light emitting diodes based upon
conjugated polymers, with quantum efficiencies attractive for
consideration in real devices, is quite new[2] and currently growing
into a topic commanding the attention of a wide variety of scientists
and engineers, in both industry and academia, the world over[3]. A
great deal of the physics, and especially the chemistry, which
governs the behavior of polymer-LEDs, occurs at the polymer
surface, or the near surface region. The details are greatly determined
by the metalic contact. Information obtained from detailed studies of
the chemical and electronic structure of conjugated polymer surfaces
and interfaces with metals, is becoming a basic ingredient in
understanding device behaviour and optimizing device performance.

In this book, we attempt to bring together in one place the results
of a relatively large number of basic studies of conjugated polymer
surfaces, as well as the 'early stages of metal–polymer interface
formation', in an attempt to produce a simple and coherent picture of
some of the unique features of these surfaces and interfaces; features
which are important in understanding and controling the performance
of polymer-based LEDs. Instead of presenting a series of detailed
chronological accounts of individual studies, we have tried to take a
more global approach, at least in part, where the nature of the

information allows, in order to make the book more comprehensible to readers from a range of different backgrounds. This 'compendium' is intended to be multi-level in content; different parts and chapters being of use to readers of different backgrounds and/or interests, as outlined below.

1.1   Background: towards molecular (based) electronics

Organic materials comprise most of the chemically identifiable materials on the earth. Despite this fact, *in*organic materials have occupied the traditional role as electronic materials in the modern information technology revolution. As the density of electronic components continues to increase, the characteristic size of the individual electronic elements (e.g., transistors) are rapidly approaching the molecular scale, ~ 10 Å. It is expected, however, that bulk-effect solid state physics devices will 'mature' at dimensions on the order of 400 atoms (i.e., around 50 to 60 nm)[4], at which point quantum-physics devices and molecular-based elements will come into play. Either individual organic molecules or molecular aggregates become attractive candidates for active elements; at first certainly in specialized electronic applications, later as major components. With only a few exceptions, molecular materials are organic, that is, based upon carbon.

Molecular electronics on a true molecular size scale is a technology with a future somewhere well into the next century[5]. On the other hand, on a larger size scale, bulk molecular materials, in thin-film form, have been in use for many years. As one example, organic photo-receptors have been at the heart of the photocopying industry for almost two decades[6]. More recent developments involve conjugated polymers[7], which have been the subjects of intense study, since, in 1977, Heeger, MacDiarmid, Shirakawa and co-workers were able to dope polyacetylene to a high electrical conductivity[8] of almost 10^{+3} S/cm. Although conjugated polymers attracted much initial interest based upon their electrical properties in the doped, electrically conducting state, new developments in device applications are based upon these polymers in the pristine, semiconducting state. Conjugated-polymer-based transistors, and especially conjugated-polymer-based light emitting diodes, or LEDs, developed at the University of

Cambridge, UK, are presently the focus of study in a large number of research laboratories, and topics of development in ever-growing numbers of both small and large industrial laboratories, world wide[9].

1.2   Surfaces and interfaces

It is perhaps trivial and obvious to state that communication with a material occurs through the surface, and that communication involving physical contact occurs through a physical interface. It also is clear that addressing the properties of a material involves understanding the nature of surfaces and interfaces of that material. Although the science of inorganic surfaces and interfaces is highly developed[10], the surface science of organic materials, and in particular polymeric materials, is less well developed[11]. Even then, most work reported to date involves the surfaces and interfaces on non-conjugated polymers. The early work was pioneered by Clark[12], while more recently the science is represented by the work of Pireaux and co-workers[13]. Some examples of applications have been documented[14] in the series of books edited by Mittal. The surface and interface science of conjugated polymers, although sparsely studied, is highly relevant to the present and rapidly developing areas of molecular electronics and molecular-based electronics[3].

1.3   Approach

This book represents an attempt to illustrate the state-of-the-art of the basic physics and materials science of conjugated polymers and their interfaces, as related more-or-less to a specific application, present polymer-based LEDs, as of the time of writing (March 1995). It is intended to illustrate the basic fundamental physical (and the specifically associated chemical) principles that apply to these materials, which in many instances are different than those encountered in their inorganic counterparts. The references provided are not exhaustive, but rather are representative of the state-of-the-art; as such, they should be sufficient to enable the interested reader to delve deeper into the area. Although by doing so some duplication occurs, references are grouped at the end of each chapter for convenience. As history has shown in other areas of physics and chemistry, it may be that some presently 'known' properties might actually depend (at least to some extent) upon the experimental measurement used to study the property, or even the theoretical model used in the interpretation of measurement data. Therefore, the goals here are to point out specific issues and specific properties of polymer surfaces and interfaces, which may be relevant in the context of the emerging area of polymer-based electronics, in general, and polymer-based light emitting diodes in particular; and, in the process, discuss both the theoretical and experimental methods mostly commonly used in studies of these properties[15]. Generally, but especially in the introductory and background material, the discussions are made as short, concise and to-the-point as possible, pointing out the features relevant to the specific topics at hand.

Although it is not intended to provide here a detailed description of polymer-based LED devices, some background material is included, both as motivation as well as to set the context for, and develop the terminology for, the discussions of the surface and interface issues. Such an approach should enable a scientist entering this field to become familiar enough with the most applicable theoretical and experimental methods employed, so that he/she should be able to read and understand the basic scientific papers, and to understand the issues which are particularly relevant to addressing the underlying science and development of polymer-based light emitting diodes. The level of discussion is aimed at the intermediate graduate student level, which also should be sufficient for the established, but non-expert, scientist, new to the area of polymer surfaces and interfaces.

1.4   Structure and organization

In addition to the present *Introduction* chapter, the contents are organized and presented in chapters, and subdivided in such a way that, to the extent reasonably possible, each chapter comprises a free-standing unit, which may be read (by readers of appropriate backgrounds) as independent units, if or when necessary.

■ *Theory and experimental methods*. Since the combined experimental–theoretical approach is stressed, both the underlying theoretical and experimental aspects receive considerable attention in chapters 2 and 3. Computational methods are presented in order to introduce the nomenclature, discuss the input into the models, and the other approximations used. Thereafter, a brief survey of possible surface science experimental techniques is provided, with a critical view towards the application of these techniques to studies of conjugated polymer surfaces and interfaces. Next, some of the relevant details of the most common, and singly most useful, measurement employed in the studies of polymer surfaces and interfaces, photoelectron spectroscopy, are pointed out, to provide the reader with a familiarity of certain concepts used in data interpretation in the *Examples* chapter (chapter 7). Finally, the use of the output of the computational modelling in interpreting experimental electronic and chemical structural data, the combined experimental–theoretical approach, is illustrated.

■ *Materials*. A detailed discussion of the chemical and electronic structure of conjugated molecules and polymers is unnecessary, since several recent and comprehensive reviews are available[16–18]. A few of the aspects essential for the chapters which follow, however, are presented in chapter 4. First, a description of the concept of molecular and certain polymeric solids, in contrast to conventional three-dimensional covalently bonded semiconductors and insulators, is provided. This is in effect necessary in order to bring the language of physics and chemistry together. Thereafter, a brief description of the basic physical properties of π-conjugated polymers is included to set the context for discussions of the elementary optical excitations, and the nature of the charge bearing species in these systems. Then, the basic electronic structure of one-dimensional (linear) conjugated polymers is outlined in terms of the electronic band structure[19]. The use of model molecular solid (thin film) systems for conjugated polymer systems is described, but only to the extent necessary for some of the examples. Finally, key issues involved in optical absorption and photo-luminescence are outlined, emphasizing the interconnection of the geometrical and electronic structure of π-conjugated systems.

■ *Device motivation for interface studies,* and *Optical absorption and emission in conjugated oligomers and polymers.* The principles of device physics of metal insulator field-effect transistors (MISFETs) and light emitting diodes (LEDs) are outlined; mainly as motivation for the contents of the chapters which follow, but also to point out certain features relevant to developing an understanding of the nature of the polymer–metal interface (chapters 5 and 6). The basic principles of electro-luminescence are reviewed here, at the level consistent with the aims of this work.

■ *Examples.* Combined experimental–theoretical studies lead to information at a level not easily obtainable from either approach separately[15]. Several detailed examples are provided in chapter 7 to illustrate this point, and to provide the basis for the conclusions drawn on relevant polymer surfaces and the early stages of metal–polymer interface formation. This portion of the book is for the reader who wants to become familiar with details upon which certain conclusions, in the final chapter, have been drawn.

■ *The nature of organic molecular solid surfaces and interfaces with metals.* This is a summary containing a digest of the results of the investigations, where 'our view of polymer surfaces and interfaces', in the context of polymer-based LEDs, is summarized in a direct way. Sub-divisions include: polymer surfaces; polymer-on-metal interfaces; and polymer–polymer interfaces. Different (ideal) models of the interfaces are outlined.

1.5 References

1. M. Pope, H. P. Kallmann and A. Magnante, *J. Chem. Phys.* **38**, 2042 (1963).

2. J. H. Burroughes, D. D. C. Bradley, A. R. Brown, R. N. Marks, K. Mackay, R. H. Friend, P. L. Burn and A. B. Holmes, *Nature* **347**, 539 (1990).

3. J. L. Brédas, W. R. Salaneck and G. Wegner (Eds), *Organic Materials for Electronics: Conjugated Polymer Interfaces with Metals and Semiconductors* (North Holland, Amsterdam, 1994).

4. J. S. Mayo, *Physics Today* **47**, (12), 51 (1994).

5. M. C. Petty, M. R. Bryce and D. Bloor (Eds), *Introduction to*

Molecular Electronics (Edward Arnold, London, 1995).

6. J. Mort, *Physics Today* **47**, (4), 32 (1994).

7. W. R. Salaneck, D. T. Clark and E. J. Samuelsen (Eds), *Science and Technology of Conducting Polymers* (Adam Hilger, Bristol, 1991).

8. C. K. Chiang, C. R. Fincher, Y. W. Park, A. J. Heeger, H. Shirakawa, E. J. Louis, S. C. Gao and A. G. MacDiarmid, *Phys. Rev. Lett.* **39**, 1098 (1977).

9. J. P. Farges (Ed.), *Organic Conductors: Fundamentals and Applications* (Marcel Dekker, New York, 1994).

10. C. B. Duke (Ed.), *Surface Science: The First Thirty Years* (North-Holland, Amsterdam, 1994).

11. R. Hoffman, *Solids and Surfaces: A Chemist's View of Bonding in Extended Structures* (VCH, New York, 1988).

12. D. T. Clark and W. J. Feast (Eds), *Polymer Surfaces* (John Wiley & Sons, Chichester, 1978).

13. J. J. Pireaux, in *Surface Characterization of Advanced Polymers*, L. Sabbatini and P. G. Zambonine (Eds) (VCH, Mannheim, 1993).

14. K. L. Mittal (Ed.), *Metallized Plastics* (Plenum, New York, 1991).

15. J. L. Brédas and W. R. Salaneck, in *Organic Electroluminescence*, D. D. C. Bradley and T. Tsutsui (Eds) (Cambridge University Press, Cambridge, 1995), in press.

16. W. R. Salaneck and J. L. Brédas, *Solid State Communications, Special Issue on 'Highlights in Condensed Matter Physics and Materials Science'* **92**, 31 (1994).

17. M. Schott and M. Nechtschein, in *Organic Conductors: Fundamentals and Applications*, J. P. Farges (Ed.) (Marcel Dekker, New York, 1994), p. 495.

18. M. Schott, in *Organic Conductors: Fundamentals and Applications*, J. P. Farges (Ed.) (Marcel Dekker, New York, 1994), p. 539.

19. W. R. Salaneck and J. L. Brédas, in *Organic Materials for Electronics: Conjugated Polymer Interfaces with Metals and Semiconductors*, J. L. Brédas, W. R. Salaneck and G. Wegner (Eds) (North Holland, Amsterdam, 1994), p. 15.

- *Chapter 2*

Theory

2.1 Introduction

The purpose of performing calculations of physical properties parallel to experimental studies is twofold. First, since calculations by necessity involve approximations, the results have to be compared with experimental data in order to test the validity of these approximations. If the comparison turns out to be favourable, the second step in the evaluation of the theoretical data is to make predictions of physical properties that are inaccessible to experimental investigations. This second step can result in new understanding of material properties and make it possible to tune these properties for specific purposes. In the context of this book, theoretical calculations are aimed at understanding of the basic interfacial chemistry of metal-conjugated polymer interfaces. This understanding should be related to structural properties such as stability of the interface and adhesion of the metallic overlayer to the polymer surface. Problems related to the electronic properties of the interface are also addressed. Such properties include, for instance, the formation of localized interfacial states, charge transfer between the metal and the polymer, and electron mobility across the interface.

In this chapter we discuss theoretical modelling, approximation schemes, and calculation methods. The description of the methods is on the level where we focus on the main steps in the theoretical development and not on details related to the solution of the resulting equations. For details of the derivation and evaluation of these equations we refer to some of the excellent books or reviews that are available[1,2]. Special emphasis is put on quantities that are subject to direct comparison with experimental data, for instance, electron binding energies and how the approximations invoked in the theory influence the accuracy of these quantities. The analysis and evaluation of theoretical results in comparison with experimental electronic structure data are discussed in chapter 7.

2.2 Schrödinger equation

The basis for studies of the electronic structure of materials is the (non-relativistic) time independent Schrödinger equation, $H\Phi = E\Phi$, where the Hamiltonian, H, in atomic units is:

$$H = -\sum_i \frac{1}{2}\nabla_i^2 - \sum_A \frac{1}{2M_A}\nabla_A^2 - \sum_{i,A} \frac{Z_A}{|r_i - R_A|} + \qquad (2.1)$$
$$\frac{1}{2}\sum_i\sum_j \frac{1}{|r_i - r_j|} + \frac{1}{2}\sum_i\sum_A \frac{Z_A Z_B}{|R_A - R_B|}$$

where r_i and R_A are the position vectors for the electrons and nuclei. The two first terms in eq. (2.1) are the operators for the kinetic energy of the electrons and the nuclei, respectively, the third term expresses the Coulomb attraction between the electrons and the nuclei, and the last two terms express the Coulomb repulsion between pairs of electrons and pairs of nuclei. The solution of the time-independent Schrödinger equations gives the stationary states of the system.

The experimental spectroscopic methods discussed below are performed in the steady state, i.e., the time average of the nuclei positions is fixed. This justifies the use of the time-independent Schrödinger equation in the calculations. Dynamical systems are also of some interest in the context of metal–polymer interfaces in studies of, for instance, the growth process of the metallic overlayer. Also, in the context of polymer or molecular electronic devices, the dynamics of electron transport, or transport of coupled electron–phonon quasi-particles (polarons) is of fundamental interest for the performance of the device. However, in the context of this book we restrict ourselves to studies of stationary states and to the situation when the interface is already formed.

The first step in solving the Schrödinger equation is to apply the so-called Born–Oppenheimer approximation. This approximation makes use of the fact that the nuclei are much heavier than the electrons and therefore move much slower. Any change in the position of a nucleus leads to an immediate response of the electronic system. The kinetic energy of the nuclei can therefore be neglected in calculations of the total energy of the system. Consequently, the total energy is expressed as a sum of an electronic part and a constant nuclear repulsion energy term (the last term in eq. (2.1) above). The electronic energy is obtained by solving the electronic Schrödinger equation $H_{el}\Psi = E_{el}\Psi$, where:

$$H_{el} = -\sum_i \frac{1}{2}\nabla_i^2 - \sum_i \sum_A \frac{Z_A}{|r_i - R_A|} + \frac{1}{2}\sum_i \sum_j \frac{1}{|r_i - r_j|} \qquad (2.2)$$

In the expression for the electronic energy the positions of the nuclei can be regarded as parameters that can be set to any value we like. In such a way one can explore the potential energy surface of the system. Another common situation is that only the equilibrium geometrical configuration is of interest. In this case the total energy of the system has to be minimized with respect to the position of the nuclei. The latter is an important part of the studies of how metal atoms interact with carbon-based organic molecules. The procedure of geometry optimization is discussed in Section 2.9 below.

If we are interested in the ground-state electronic properties of a molecule or solid with a given set of nuclear coordinates we should seek the solution to the Schrödinger equation which corresponds to the lowest electronic energy of the system. However, the inter-electronic interactions in eq. (2.2) are such that this differential equation is non-separable. It is therefore impossible to obtain the exact solution to the full many-body problem. In order to proceed, it is necessary to introduce approximation in this equation. Two types of approximations can be separated, namely, approximations of the wavefunction, Ψ, from a true many-particle wavefunction to, in most cases, products of single particle wave functions (orbitals) and approximations of the Hamiltonian. In fact, many theoretical models involve both of these approximations. Sections 2.3–2.8 below present the most commonly used approximation schemes, the advantages and drawbacks of the different schemes, the physical significance of the results and their applicability to calculations of metal–polymer interfaces.

2.3   Hartree–Fock theory

Ab initio Hartree-Fock theory is based on one single approximation, namely, the *N-electron* wavefunction, Ψ is restricted to an antisymmetrized product, a Slater determinant, of one-electron wavefunction, so called spin orbitals, χ_i:

$$\Psi(1, 2, ..., N) = (N!)^{\frac{1}{2}} \sum_{N=1}^{N!} (-1)^{P_n} P_n\{\chi_i(1)\chi_j(2) \ldots \chi_k(N)\} \qquad (2.3)$$

Using this wavefunction the total energy of the electron system is *minimized with respect to the choice of spin-orbitals under the constraint that the spin orbitals are orthogonal.* The variational procedure is applied to this minimization problem and the result is the so called Hartree–Fock equations:

$$h(1)\chi_i(1) + \sum_{j \neq i} \int \frac{|\chi_i(2)|^2}{|r_1 - r_2|} dr_2 \chi_i(1) - \sum_{j \neq i} \int \frac{\chi_j(2)\chi_i(2)}{|r_1 - r_2|} dr_2 \chi_j(1) = \varepsilon_i \chi_i(1)$$

$$(2.4)$$

where the indices i and j run over the number of electrons, i.e., $i,j = 1,2,....N$ and $h(1)$ is defined by:

$$h(1) = -\frac{1}{2}\nabla^2 - \sum_A \frac{Z_A}{|r_1 - R_A|} \qquad (2.5)$$

This is the single-electron operator including the electron kinetic energy and the potential energy for attraction to the nuclei (for convenience, the single electron is indexed as electron one). The two-electron operators in eq. (2.4) are defined as the Coulomb, J

(second term) and exchange, K (third term), operators. By definition, eq. (2.4) can now be written in a more compact form as:

$$\left(h(1) + \sum_{j \neq i} J_j + K_j \right) \chi_i(1) = \varepsilon_i \chi_i(1) \qquad (2.6)$$

The effective single-particle operator in eq. (2.6) is referred to as the Fock operator, $f(1)$, i.e., eq. (2.6) can now be written as $f(1)\chi_i(1) = \varepsilon_i \chi_i(1)$.

The spin orbitals can be separated into a spatial part, ψ (orbital or molecular orbital) and a spin eigenfunction, α (or β for the opposite spin). In restricted Hartee–Fock theory (RHF), the spatial part is independent on the spin state, in contrast to unrestricted Hartree–Fock (UHF) where it is spin dependent. Consequently, the RHF spin orbitals can be written as $\chi_i = \psi_i \alpha$, whereas in UHF the corresponding relation is $\chi_i = \psi_{\alpha i} \alpha$. The discussion here is limited to RHF case and to the case of an even number of electrons (closed shell system), but can easily be extended to treat also UHF[1].

By integrating out the spin-coordinate of eq. (2.6) it simplifies to the following RHF equation:

$$\left(h(1) + \sum_{j=1}^{N/2} 2J_j + K_j \right) \psi_i(1) = \varepsilon_i \psi_i(1) \qquad (2.7)$$

Given the wavefunction of the form shown in eq. (2.3), the solution to eq. (2.7) gives the orbitals that minimize the total energy. The HF-equations are here written in the so called canonical form[1]. This form is particularly convenient since the eigenenergies, ε_i, which were introduced in the form of Lagrange multipliers in the minimization procedure, represent the binding energy (ionization potential) of individual electrons, which is exactly the quantity measured in photoelectron spectroscopy. Therefore, under the assumption that the Slater determinant wavefunction is a valid approximation, there is a very clear connection between theory and experiment as concerns the eigenvalues obtained from eq. (2.7) above.

It should be noted here that the HF approximation is not valid for metallic systems, i.e., a system with partially filled bands[2,3]. The reason for this is that the lowest unoccupied states in the HF eigenvalue spectrum do not represent eigenenergies above the Fermi energy but rather electron affinities. Since there is an extra energy term associated with bringing additional electrons into the system, there is always a gap at the Fermi energy in the HF eigenvalue spectrum in disagreement with the electronic structure of metals[4]. Another drawback of the HF method is that the Coulomb and exchange interactions are long range. Calculations on infinite systems therefore face the difficulty in evaluating lattice summations of the Coulomb and exchange integrals (the expectation value of the Coulomb and exchange operators) to satisfactory convergence. Attempts have been made to handle this problem, the lattice summations of Coulomb integrals can be computed via multipole expansion whereas the exchange lattice summations are still treated on a trial and error basis[5]. The difficulties in formulating the HF theory for crystals has lead to very few applications of this method to crystal calculations[5,6]

The great success of HF theory is in its application to molecules and clusters. It has become the standard method in the field of quantum physics known as quantum chemistry. The tremendous success of the HF method in this field is not only due to the Hartree–Fock solutions themselves but also to the fact that HF serves as a natural starting point for calculations including electron correlation. As stated above, the only approximation involved in the theory is that of expressing the wavefunction as a single Slater determinant of orbital wavefunctions. The correlation energy is therefore, by definition, the difference between the exact energy and the Hartree–Fock energy. There exists a whole range of methods dealing with the problem of electron correlation in relation to HF theory: configuration interaction, perturbation theory, coupled clusters etc. These methods are very computation intensive and have not yet become applicable to large systems unless these systems are highly symmetric. In particular in the search for equilibrium geometries of large systems, most HF calculations still have to be performed using a single determinant wavefunction. However, with the rapid development in computer power we see today, we expect that calculations involving correlation effects will very soon become the standard method to treat systems of the size discussed in this book.

2.4   Roothaan equations

We now turn to the problem of how the Hartree–Fock equations are solved in practice. The standard way to proceed is to expand the wavefunction in a basis. The orbital is expanded in terms of a linear combination of basis functions:

$$\psi_i(r) \cong \sum_{\mu=1}^{K} C_{\mu i}\phi_\mu \qquad (2.8)$$

where $C_{\mu i}$ are the expansion coefficients and ϕ_μ the basis function. Usually, the basis is chosen to consist of atomic orbitals (AO) of s,p,d, or f type. The expansion in eq. (2.8) is therefore termed linear combination of atomic orbitals (LCAO). Note that the expansion is exact only if the basis set is complete, i.e., the number of basis functions, K, is infinite. For practical reasons, however, one is restricted to a finite number of basis functions. Naturally, the accuracy of the solution depends on the choice of basis set. Qualitatively correct results are, however, obtained even for rather restricted basis sets. Calculations on large systems, of the type that are discussed in this book are therefore not subject to any major error because of the choice of a practically manageable basis set.

The normal choice of LCAO basis functions are Gaussian functions, or for larger systems, contractions of Gaussian functions. The main reason for this choice is that integrals that have to be calculated in order to solve the HF equations (see eqs. (2.9) and (2.10) below) can be calculated analytically if Gaussian functions are used. For computational purposes this is absolutely essential, especially for systems involving a large number of basis functions, since numerical evaluation of integrals is very tedious.

The formulation of the wavefunction in a basis leads to the so-called Roothaan equations which are obtained from the HF equations by inserting the expansion (2.7) into (2.4), multiplying from the left by the set of basis functions $\{\phi_v^*\}$ and integrating over the spatial coordinates. This procedure leads to the following integral matrix equation.

$$\boldsymbol{FC} = \boldsymbol{SC}\varepsilon \qquad (2.9)$$

where ε is a vector containing the K energy eigenvalues, $\boldsymbol{C}$ is a $K \times K$

matrix containing the expansion coefficients, $C_{\mu i}$ (see eq. (2.8)), which minimizes the total energy, and F and S are the Fock and overlap matrices, respectively, with elements defined by:

$$F_{\mu\nu} = \int \phi_\mu(1) f(1) \phi_\nu(1) \mathrm{d}r \qquad (2.10)$$

and

$$S_{\mu\nu} = \int \phi_\mu(1) \phi_\nu(1) \mathrm{d}r \qquad (2.11)$$

Note that the Fock operator, $f(1)$, also involves the basis functions via the spin orbitals appearing in the Coulomb and exchange operators (see eq. (2.4)). The Roothan equations (as well as the HF equations) are nonlinear and must therefore be solved iteratively by first guessing a starting electronic wavefunction, and from that calculating the Coulomb and exchange integrals and the Fock matrix. New expansion coefficients are then obtained from the solution of eq. (2.9) that define a new wavefunction, which is used in the next iteration step. This procedure is repeated until convergence is obtained, i.e., until the difference in the expansion coefficients between two successive iterations is smaller than some predefined threshold value. The solution has then reached self-consistency and the potential in which the single electron is moving is referred to as the self-consistent field (SCF). In fact, the SCF procedure is common for all methods discussed in this chapter except the VEH method.

It should be noted that in the way the Hartree–Fock theory is formulated, the number of integrals that has to be calculated increases rapidly with the size of the system. By inserting $f(1)$ from eq. (2.7) into eq. (2.9), the two-electron integrals, i.e., the Coulomb and exchange integrals, are obtained in the following form:

$$(\mu\nu \,|\, \lambda\sigma) = \int \phi_\mu(1) \phi_\nu(1) r_{12}^{-1} \phi_\lambda(2) \phi_\sigma(2) \mathrm{d}r_1 \mathrm{d}r_2 \qquad (2.12)$$

where $r_{12} = |r_1 - r_2|$. This integral is referred to as a four-centre integral. If all such four-centre integrals are calculated the number of integrals scales as $K^4/8$, where K is the number of basis functions. A lot of effort is today put into the problem of reducing the number of integrals by introducing various cut off (see also note on crystals above) and other types of approximations. The size dependence is, however, a severe limitation for calculation on larger systems. In our studies of metal–polymer interfaces we have introduced the concept of model molecules. To work with model molecules is convenient both from an experimental and theoretical point of view. Experimentally, the model system is more well defined and easier to control. Theoretically, it has the advantage of being accessible to accurate quantum chemical calculations. As a result of the possibility to perform accurate calculations, theoretical and experimental results usually agree very well, which contributes to the understanding, not only of the model molecules but also of the corresponding polymers.

2.5 Pseudopotentials and atomic operators

The basic idea behind the pseudopotential method is to treat the valence electron as moving in a potential from a fixed ion core. Chemically this is very reasonable and it is only if one is interested in very accurate investigation of spectroscopic data of elements with highly polarizable cores (typically alkali metals) that this approach fails[7].

The Fock operator defined in eq. (2.7) above, here rewritten for a closed shell system, can be diagonalized and expressed in terms of the core *(c)* valence *(v)* and excited state *(e)* projectors as[7]:

$$f(1) = h(1) + \sum_j (2J_j - K_j) = \tag{2.13}$$

$$\sum_c \varepsilon_c |\psi_c\rangle\langle\psi_c| + \sum_v \varepsilon_v |\psi_v\rangle\langle\psi_v| + \sum_e \varepsilon_e |\psi_e\rangle\langle\psi_e|$$

The valence pseudo-Fock operator takes the form:

$$f^{ps} = h(1) + V^{ps} + \sum_v (2J_v - K_v) = \tag{2.14}$$

$$\sum_v \varepsilon'_v |\psi'_v\rangle\langle\psi'_v| + \sum_e \varepsilon'_e |\psi'_e\rangle\langle\psi'_e|$$

The pseudopotential, V^{ps}, may be derived by minimizing the distance between f^{ps} and either the full Fock operator, i.e., $\|f^{ps} - f\|_{min}$, or the truncated valence Fock-like operator, $\|f^{ps} - f_v\|_{min}$, where $f_v = \sum_v \varepsilon_v |\psi_v\rangle\langle\psi_v|$. The use of pseudopotentials for transition metal atoms is volumous and successful. The method is applied, for instance, in studies of molecules on metallic surfaces[8]. The metallic surface is approximated by a large cluster of metal atoms. The cluster can be made fairly large if the inner electrons are described by a pseudopotential and only the included outer valence electrons are described quantum mechanically. Of course, the accuracy of the method depends on which electrons are treated explicitly. For copper, for instance, one may consider the 3s3p3d4s shells explicitly, or the 3d4s shells, or only the 4s shell. In our studies we have so far not used the *ab initio* pseudopotential technique since the number of heavy atoms that we have treated has been rather restricted.

However, this is definitely the technique for future calculations involving a large number of metal atoms. Furthermore, the idea behind the pseudopotential method is also applied in other types of Hamiltonians described below, e.g., valence effective Hamiltonian and semi-empirical methods.

The concept of atomic operators is related to the question whether or not it is possible to define universal purely monoelectronic potentials characteristic of specific atoms. By universal is meant that the same atomic potential can be applied to a specific atom, carbon say, for a whole class of carbon-containing molecules. Potentials of this type should not include explicitly the wavefunctions of the other electrons in the system, i.e., *this approach is not subject to the iterative SCF solution method discussed above in connection with the HF theory.* Calculations based on atomic operators are therefore rather fast.

Consider a molecule built up from different types of atoms A. It is assumed that the monoelectronic operator can be written as a sum of the kinetic energy and various atomic potentials:

$$f^{ps} = -\frac{1}{2}\nabla^2 + \sum_A V_A{}^{ps} \qquad (2.15)$$

where $V_A{}^{ps}$ is the monoelectronic (pseudo)potential for atom A. This potential can be expressed in operator form as (c.f. eq (2.13) above):

$$V_A^{ps} = \sum_{pq} C_p{}^q |p\rangle\langle q| \qquad (2.16)$$

In the treatment of Nicolas and Durand[9], the projectors $|p\rangle$ are expressed in terms of Gaussian functions. To determine the coefficients C_{pq}, one performs the same type of minimization as in the case of real pseudopotentials, i.e., the monoelectron operators should be defined in such a way that they simulate as close as possible the exact Fock operator. In order to make the atomic potential as general as possible, a set of molecules (M1,M2,...,Mn) is selected and the sum for these molecules is minimized: $(\|f^{ps} - f\|_{M_1} + \|f^{ps} - f\|_{M_2} + \ldots + \|f^{ps} - f\|_{Mn})_{min}$. Atomic potentials for hydrogen, carbon, nitrogen, oxygen, sulphur, silicon are determined for this type of Hamiltonian under the name of valence effective Hamiltonian, or VEH. It is notable that the valence effective Fock operator does not involve the Coulomb and exchange operators. The difficulties involved in calculating the expectation values of these operators for extended systems due to their long range nature (see section 2.2 above) do not arise for the operator in eq. (2.15). On the contrary, the Gaussian functions used to express these effective operators give excellent convergence properties of the two-centre integrals at large separations of the centra. The valence effective Hamiltonian is therefore very suitable for crystal calculations. In particular, it has been extended to treat polymers, i.e., systems that have translational symmetry in one dimension only. Details about this approach is given in section 2.7 below.

2.6 Semi-empirical methods

In order to make calculations on large systems possible, e.g. optimization of inter-atomic distances in the types of system that are of interest for metal–polymer interfaces, it is often necessary to work with more approximate theoretical methods. This does not automatically leads to poorer results than from, for instance, a minimal basis set *ab initio* Hartree–Fock calculation. However, since the method relies on a kind of fitting procedure it is particularly important to make comparison with experimental data since this is the only test of whether this procedure is valid or not.

The approximation schemes that are discussed in this paragraph are all based on the Hartree–Fock theory, i.e., the many-electron wavefunction is described in terms of a single Slater determinant.

Instead of approximating the wavefunction further, the focus is now on the Hamiltonian. In particular, as discussed above, the two-electron integrals (see eq. (2.11)) are responsible for the rapid increase in both computational time and memory requirements. The idea is here to avoid calculating integrals that are small and thus are relatively unimportant for the results. Furthermore, in combination with the integral approximations, a parametrization of the Hamiltonian is also introduced; in particular, a complete parametrization of the core electrons is performed. This approach is similar to the pseudopotential method described above. However, in this case the parameters that appear in the pseudopotentials are determined by minimizing the difference between calculated data and measured data. Therefore, this approach is referred to as semi-empirical. The minimization is done for a selection of test molecules in order to make the empirical potential as general as possible. The minimization is also done for specific physical properties, for instance the valence band photoelectron spectra.

The semi-empirical approach was first formulated by Pople[10]. In the first version, the approximation made on the two-electron integrals (see eq. (2. 12))was of the form:

$$(\mu\nu \mid \lambda\sigma) = (\mu\mu \mid \lambda\lambda)\delta_{\mu\nu}\delta_{\lambda\sigma} \qquad (2.17)$$

Effectively, this approximation puts all integrals involving products of atomic orbitals referring to the same electron to zero unless the two orbitals are identical. This approximation is referred to as complete neglect of differential overlap and gave the name to the method CNDO. Associated with the integral approximation above, there is a set of experimental parameters that enter the mono-electronic part of the Hamiltonian, $h(1)$ in eq. (2.7) above as well as the integral $(\mu\mu \mid \lambda\lambda)$ itself. As a result of this type of approximation, there is no need to express the basis in terms of Gaussian functions as in the *ab initio* treatment. In fact, all integrals are evaluated using parametrized analytical expressions and the basis orbitals are therefore never used explicitly.

Following the initial work of Pople, there appeared a number of integral approximation schemes, all combined with different sets of empirical parameters. One such approximation scheme that has been shown to be particularly useful is the neglect of diatomic differential overlap, NDDO[11]. Here, differential overlap between atomic orbitals is neglected only if the orbitals belong to different sites (atoms). Integrals of the type $(s_A p_A | s_B p_B)$ where A and B refer to two different atoms are therefore included at the NDDO level. Because of the different symmetry of the s- and p-orbitals these integrals decay as R_{AB}^{-3}. In the CNDO approximation these integrals are completely neglected. In particular, it has been shown that the short-range R_{AB}^{-3} interactions are important for optimization of interatomic separations. The NDDO level of approximation is therefore applied in methods involving geometry optimization. A whole range of such methods exist today, MINDO[12], MNDO[13] AM1[14] and PM3[15]. These methods have been shown to work very well to predict ground state geometries of hydrocarbons, hydrocarbons with heteroatoms (S,N) and also for organo-aluminum complexes and are extensively applied to our studies of metal–polymer interfaces. The capacity of modern workstations allows for calculations at the AM1 or PM3 levels on systems of several hundred heavy atoms (e.g. carbon). In the search for equilibrium geometries of clusters, the need for further approximations is therefore limited.

2.7   Polymers treated with the valence effective Hamiltonian

For the case of stereo-regular conjugated polymers, the essential step involves extending the molecular orbital eigenvalue problem to systems with one-dimensional periodic boundary conditions, i.e., over regularly repeating monomeric units. Of course, the polymer solid is really three-dimensional but the weak, van der Waals type of interactions between the polymer chains can be neglected for the purpose of this work. The translational symmetry of the polymer implies the following property of the polymer crystal orbitals:

$$\psi\,(k,r) = u_k(r)\mathrm{e}^{ikx} \tag{2.18}$$

where $u_k(r)$ has the periodicity of the lattice, x is the projection of r on the crystal axis, which is chosen to be the x-axis, and k is the crystal momentum. The periodicity in the reciprocal space implies that $\psi(k,r) = \psi(k + K,r)$ where K is a reciprocal (one-dimensional) lattice vector. The first Brillouin zone (BZ) is defined by the region between $-\pi/a < k < +\pi/a$ where a is the real space lattice constant.

Using the same type of LCAO expansion as for molecules (see eq. (2.8)), the crystal orbitals are expanded as Bloch sums of the basis function centred at site μ in cell j:

$$\psi(k, r) = \frac{1}{\sqrt{2N + 1}} \sum_{j=-N}^{N} e^{ijka} \sum_{\mu} \phi_{\mu}(r - R_{\mu} - ja)C_{\mu}(k) \quad (2.19)$$

$2N + 1$ is the number of unit cells of the polymer chain. The LCAO expansion coefficient matrices $C(k)$ and the corresponding energy eigenvalues, $\varepsilon(k)$, are obtained by solving the eigenvalue problem for each value of k:

$$F(k)C(k) = S(k)C(k)\varepsilon E(k) \quad (2.20)$$

Similar to eq. (2.9), F and S are the effective Fock and overlap matrices, respectively. In this case the matrix elements are lattice sums over all unit cells in the crystal:

$$F_{\mu\nu}(k) = \sum_{j=-N}^{N} e^{ijka} \int \phi_{\mu}(r) f^{VEH} \phi_{\nu}(r - R_{\nu} - ja)dr \quad (2.21)$$

$$S_{\mu\nu}(k) = \sum_{j=-N}^{N} e^{ijka} \int \phi_{\mu}(r) \phi_{\nu}(r - R_{\nu} - ja)dr \quad (2.22)$$

The one-electron valence effective Fock operator, f^{VEH} is that introduced in section 2.5, eq. (2.15), but extended by summing over all unit cells in the polymer[16]:

$$f^{VEH} = -\frac{1}{2}\nabla^2 + \sum_{j}\sum_{A} V_A(j) \quad (2.23)$$

The energy eigenvalues obtained from eq. (2.19) give the electronic band structure of the polymer. The set of eigenvalues that are obtained for each value of k belong to different bands. Like the crystal orbitals, the band structure is a periodic function of k. It is therefore enough to consider the first BZ only. (The one-dimensional energy band of *trans*-polyacetylene is illustrated in Fig. 3.1. The energy bands are occupied up to the Fermi energy and empty above this energy.)

The most directly measurable feature of the electronic band structure, the valence band photoelectron spectrum, involves the density-of-valence-states (DOVS), or $\rho(\varepsilon)$, which is obtained from the electronic band structure using the standard definition,

$$\rho(\varepsilon) = \left(\frac{d\varepsilon}{dk}\right)^{-1} \qquad (2.24)$$

The magnitude of $\rho(\varepsilon)$ corresponds to the number of energy states (allowed k-points) per unit energy. Maxima (peaks) in $\rho(\varepsilon)$ correspond to portions of the band structure which are relatively flat. (In the example in Fig. 3.1, the relationship between the band structure and the $\rho(\varepsilon)$ (or DOVS), for *trans*-polyacetylene is presented.)

2.8   Density functional theory

Density functional theory (DFT) and its predecessors, has been as fundamental in solid state physics as HF theory in quantum chemistry. DFT is today also becoming more and more popular for predicting ground-state properties and disassociation energies of molecules. The method has its advantages as compared to the HF method in that it, to some extent, includes electron correlation effects. It also works well for crystal calculations, in particular on metallic systems, and it becomes less computationally demanding for large systems since there are no four-centre integrals to be calculated. Unlike HF theory, in which the wavefunction plays the central role in approximating the interacting electron problem, DFT reduces the Hamiltonian of a system of interacting electrons to a single-particle equation. The price paid for this simplification is that the so-called exchange-correlation part of the single-particle Hamiltonian is unknown and has to be approximated.

The basis for DFT was given by Hohenberg and Kohn[17] who showed that the ground-state energy, E, of a system of N interacting electrons in an external field (e.g. the field from the nuclei) is determined uniquely by the electron density, $\rho(r)$, i.e., the ground-state energy is a functional of the electron density: $E = E[\rho]$. Furthermore, the exact ground state of the system, $E_{gs}[\rho_{gs}]$, corresponds to the minimum value of $E = E[\rho]$. The ground-state energy is therefore obtained by minimizing $E[\rho]$. Kohn and Sham[18] introduced a scheme for this minimization in which the energy functional is partitioned according to:

$$E[\rho] = T_0[\rho] + \int \rho(r)v_{ext}(r)dr + \frac{1}{2}\int\frac{\rho(r)\rho(r')}{|r - r'|}drdr' + E_{xc}[\rho] \quad (2.25)$$

where T_0 is the kinetic energy the system would have had if the particles were non-interacting, The advantage with this separation is that T_0 can be evaluated with arbitrary numerical accuracy. The second and third terms on the right-hand side of eq. (2.25) are the external potential energy and the Coulomb repulsion energy, respectively. Also these quantities can in principle be calculated exactly. The last term is, by definition, the exchange-correlation energy which includes essentially all interactions that are not included in the other terms.

The ground-state energy and charge *density* are obtained by minimizing the energy functional with the constraint that the number of particles is conserved. This minimization results in the following Euler equation[18]:

$$\frac{\delta T_0[\rho]}{\delta\rho(r)} + v_{ext}(r) + \int\frac{\rho(r')}{|r - r'|}dr' + \frac{\delta E_{xc}[\rho(r)]}{\delta\rho(r)} = 0 \quad (2.26)$$

If the same minimization procedure is applied to a system of non-interacting particles moving in an effective potential, V_{eff}, the corresponding Euler equation would be:

$$\frac{\delta T_0[\rho(r)]}{\delta\rho(r)} + V_{eff} = 0 \quad (2.27)$$

If we identify the last three terms of eq. (2.26) with V_{eff}, then clearly the same Euler equations will result since T_0 was defined as the kinetic energy of a system of non-interacting particles. For non-interacting particles, the Schrödinger equation can be written simply as: $H\psi_i = \varepsilon_i\psi_i$, which in the case of the DFT Hamiltonian results in the following single particle Schrödinger equation:

$$\left(-\frac{1}{2}\nabla^2 + v_{ext}(r) + \int\frac{\rho(r')}{|r - r'|}dr' + v_{xc}(\rho(r))\right)\psi_i = \varepsilon_i\psi_i \quad (2.28)$$

where

$$v_{xc}(\rho(r)) = \frac{\delta E_{xc}[\rho(r)]}{\delta\rho(r)} \quad (2.29)$$

and

$$\rho(r) = \sum_i|\psi_i|^2 \quad (2.30)$$

The crucial simplification in the density functional scheme is the relationship between the system of interacting particles, whose energy and particle density we seek, and the fictitious, non-interacting system which is actually solved. Obviously, the success of this simplification lies in the possibility to obtain an adequate expression for the exchange-correlation energy. The exchange-correlation energy of the interacting system can be expressed exactly as:

$$E_{xc}[\rho(r)] = \frac{1}{2}\int\rho(r)\left(\int\frac{\rho(r,r'-r)}{|r-r'|}dr'\right)dr \quad (2.31)$$

Kohn and Sham[18] suggested a local density approximation (LDA) for the second integral:

$$E_{xc}{}^{LDA}[\rho(r)] = \frac{1}{2}\int\rho(r)\varepsilon_{xc}(\rho(r))dr \quad (2.32)$$

where ε_{xc} is the exchange-correlation energy per particle in a uniform electron gas of density ρ. To improve this approximation, a spin polarized version of the exchange-correlation energy was later introduced[19]:

$$E_{xc}^{LSD}[\rho(r)] = \frac{1}{2}\int\rho(r)\varepsilon_{xc}(\rho_\alpha(r),\rho_\beta(r))dr \quad (2.33)$$

where α and β refer to different spins. In particular for studies of the energy of chemical reactions, but also, as we will see below, for ground-state geometries, it is necessary to go beyond the local approximation. Today, the standard way to do that is by using so-called gradient corrections, i.e., the non-local effects are approximated with a linear expansion of the charge density around the local site[20].

$$E_{xc}^{DFTG} = E_{xc}^{LSD} + E_x^G + E_c^G \qquad (2.34)$$

where E_x^G and E_c^G are the non-local or gradient corrections to the exchange and correlation energies, respectively. Several forms of these corrections are discussed. With the increase in accuracy using these corrections, DFT has received a dramatic increased attention also in the quantum chemical community[21]. It should be emphasized that the LCAO expansions of the orbital wavefunctions (eq. (2.7)), the way to optimize equilibrium geometries (see next section), and performing population analysis (see section 2.11 on population analysis below) is more or less the same in DFT as in HF. Some difficulties arise using DFT to optimized geometries because of the exchange-correlation potential, which normally is obtained from a numerical integration[22]. The similarities between the two theories has also led to a systematic comparison of these theories. Such a comparison is extremely useful in that it provides guidelines for the choice of method for studies of a specific molecule or class of molecules.

2.9   Geometry optimization

Computationally, a large part of the work related to the subject of this book has involved determination of equilibrium geometries of metal–polymer complexes and the relation between geometrical and electronic structure. Since there is a lack of experimental information concerning equilibrium geometries, these data are only accessible via calculations in which the total energy is minimized with respect to the positions of the nuclei.

In this paragraph we describe shortly the main steps involved in optimization of equilibrium geometries. The optimization is based on the methods for total energy calculations discussed above (*ab initio* Hartree–Fock, semi-empirical methods, density functional theories).

We do not directly address the methods based completely on empirical force fields, so-called molecular mechanics calculations. Such calculations are applied to very large molecules but with energy functions that can be calculated very quickly as compared to quantum mechanical methods. The methods for geometry optimization at this level of theory are therefore somewhat different from those used in the context of detailed total energy calculations.

The concept of equilibrium geometries and energy surfaces originates from the Born–Oppenheimer approximation (see above). The minima in the total energy surface can then be identified with the classical picture of a certain geometric configuration of the nuclei. If the Born–Oppenheimer approximation is not valid, for example in the vicinity of surface crossings, non-adiabatic effects are important and the meaning of a classical chemical structure is less clear. This situation is less relevant to our studies of metal–polymer interfaces and will not be discussed here.

An extremum in the energy surface is characterized by the first and second derivatives of the total energy with respect to all linearly independent translational degrees of freedom of the atoms in the molecule. The first derivative is the negative of the force acting on the atom in the direction of the variation. The second derivatives form a matrix called the Hessian with elements equal to the force constants. Clearly, in order for a structure to be stationary, the force acting on the atoms must be zero. The type of stationary point is determined by the eigenvalues of the Hessian. If all these eigenvalues are positive, the stationary point is at an energy minimum. Thus, the geometry optimization procedure must in principle include calculation of both these derivatives.

In *ab initio* Hartree–Fock theory, the basis functions usually consist of contracted Gaussian functions. This choice of basis makes it possible to calculate the electronic integrals and the integral derivatives analytically. Since the coefficients, i.e., the C-matrix in eq. (2.9) are determined variationally in such a way that a change in the coefficients does not cause any change in the total energy (see above), the derivatives of the coefficients are therefore zero. To determine the first derivative of the total energy, only the integral derivatives are needed. Since these can be evaluated analytically, the first derivatives are also obtained analytically and with little computational effort. To obtain the second derivative of the total energy, the second derivative of the integrals and the first derivative of the coefficients must be calculated. Methods for calculating these derivatives are developed and incorporated in most modern *ab initio* HF program packages[22–24]. However, in general, the second derivatives of the total energy are both costly to calculate and require a large memory to store. Therefore, there are other methods to obtain approximations of the Hessian that are more cost effective. For details on how this is done we refer to one of the recent reviews on this subject[25].

2.10   Chemical bonds and population analysis

Most metals of interest in the context of polymer-based electronic devices form some kind of chemical bond to the polymer upon interaction with a polymer surface. Population analysis, based on the electronic structure, is used to determine the character of this bond. According to the commonly used chemical terminology, bonds are classified as ionic if the bonded atoms are oppositely charged and held together by the attractive Coulomb force, and covalent if the two atoms are neutral but share the same pair of electrons. In the latter case, much of the electron density is located between the bonded atoms whereas for the ionic bond the charge density is concentrated at the atomic sites.

The degree of ionicity in the bond between a metal atom and a polymer, or molecule, is related to the ionization potential and electron affinities of the substituents. The metals we have studied are of interest as electron injecting contacts in electronic devices. These metals must have a low ionization potential (or work function), of the same order as the electron affinity of the polymer, in order for the charge transfer process to occur. If the ionization potential of the metal is lower than the polymer–electron affinity, spontaneous charge transfer occurs which is the signature of an ionic bond. Thus, the character of the charge distribution in the metal–polymer complexes we are studying is related to the situation in the electronic device.

In the case of the HF method, the population analysis is based on the SCF density matrix which is obtained from the molecular orbital coefficients (eq. (2.8)) in the following way:

$$P_{\mu v} = \sum_{i}^{occ} C_{iv} C_{i\mu} \qquad (2.35)$$

The summation over i includes all occupied levels of the energy eigenvalue spectrum. In the case of a closed shell system this number equals $N/2$, where N is the number of electrons in the system. Using eqs. (2.8) and (2.35), the electronic charge density can be expressed as:

$$\rho(r) = \sum_{i}^{occ} |\psi_i|^2 = \sum_{\mu} \sum_{v} P_{\mu v} \phi_\mu(r) \phi_v(r) \qquad (2.36)$$

$\rho(r)$ represents the probability of finding an electron at position r. By integrating eq. (2.36) over all space we get, by definition, the total number of electrons

$$N = \int \rho(r) dr = \sum_{\mu} \sum_{v} P_{\mu v} \int \phi_\mu(r) \phi_v(r) dr \qquad (2.37)$$

Using the definition of the overlap matrix (see eq. (2.11) above) we can rewrite this as:

$$N = \sum_{\mu} \sum_{v} P_{\mu v} S_{v\mu} = \sum_{\mu} (PS)_{\mu\mu} \qquad (2.38)$$

There is no unique definition of the number of electrons to be associated with a given nucleus in a molecule[1]. However, given eq. (2.38), we find one obvious possibility, i.e., to interpret $(PS)_{\mu\mu}$ as the number of electrons to be associated with the basis function ϕ_μ. Assuming that the basis functions are centred on atomic nuclei, as in the case of an LCAO expansion (see eq. (2.8)), the corresponding number of electrons to be associated with a given atom in a molecule is obtained by summing over all basis functions centred on that atom. The net charge associated with an atom is then given by:

$$q_A = Z_A + \sum_{\mu \in A} (PS)_{\mu\mu} \qquad (2.39)$$

where Z_A is the proton charge of atom A. This is called the Mulliken population analysis[26]. It turns out that this way to partition the electronic charge into atomic charges works well for moderate charge transfers, i.e., when the net atomic charge is small.

In order to characterize correctly the charge transfer in an ionic bond, we have used the so called natural population analysis instead[27]. This population analysis is based on expression (2.39) above, but instead of using the normal atomic orbital (AO) basis set, a set of natural atomic orbitals (NAO) are used as basis functions. The NAOs are the orthonormal atomic orbitals of maximum occupancy for the given wavefunction[27]. These orbitals are obtained following an elaborate orthogonalization procedure of the non-orthogonal AO basis. A comparison between the Mulliken charges and the natural charges for the charge transfer complex we have studied showed large differences[28]. The Mulliken charges normally lie between +0.5 and +0.7 for sodium in polyene–sodium complexes whereas the natural charges for the same compounds indicate an almost complete charge transfer, +0.9 on sodium. Experimental data clearly indicate a charge transfer which is close to unity for this type of system. For less polar compounds, however, the two different population analyses seem to give almost the same result[27].

2.11   ΔSCF

X-ray photoelectron spectroscopy of atomic core levels (XPS or ESCA) is a very powerful tool for characterization of the chemical surrounding of atoms in molecules. In particular, since the method is very surface sensitive, it is possible to monitor the first stages of the interface formation, i.e., in our case the interaction between individual metal atoms and the polymer. Standard core level bonding energies are well known for common materials. However, in our case, we are studying new combinations of atoms and new types of structures for which there are no reference data available. In order to interpret the experimental chemical shifts it is useful to compare with theoretical estimates of the shifts.

Given the HF single determinant wavefunction of an N-electron system, the Koopmans theorem[29] states that the energy required to produce an $(N\text{-}1)$-single determinant wavefunction by removing an electron from orbital ψ_i is $-\varepsilon_i$. (see eq. (2.7)). Application of Koopmans theorem to estimate the core level binding energies is subject to one major approximation, namely, that of neglect of relaxation effects in the final ionic state. The electron distribution is the same in the initial and final states except for the particular electronic state which is depopulated in the photoionization process. If we are mainly focused on chemical shifts, this error can be approximately cancelled by subtracting two calculated binding energies of the same atom in different environments. This method is, however, not very reliable. Instead, the standard way to calculate core level binding energies and chemical shifts is to perform so called delta-self-consistent field (ΔSCF) calculations[30].

The ΔSCF method includes two total energy calculations, one calculation to obtain the ground-state electronic structure of the molecule, i.e., the initial state in the photoionization process and a second calculation on the ionized final state with the hole in the core level. The binding energy of the core electrons is the difference between these two total energies. The crucial part in this type of calculations is to describe the ionized state correctly. Since the HF theory is variational, a *straightforward* calculation of the ionized system will lead to a so called variational collapse where the empty core hole is filled with electrons initially at higher energies. The variational collapse is avoided by first considering the core hole as frozen. This step takes the wavefunction to a local region of the final core hole state. In a second step, the Newton–Raphson method is used to ensure convergence to this particular final state and not to the global energy minimum[31]. The method has been applied to organic systems with an accuracy of around 1 eV for binding energies and a few tenths of an electronvolt for chemical shifts. Recent calculations on Al-polythiophene complexes have shown a very good agreement with experimental data[32]. Future powerful computers will make this type of calculation important in the interpretation of photo-emission core level spectra even on rather extended systems such as those large metal–polymer clusters that are the subject of this book.

2.12 References

1. A. Szabo and Ostlund, *Modern Quantum Chemistry* (Clarendon Press, Oxford, 1988).

2. *Ab initio Methods in Quantum Chemistry*, I. Prigogine and S. A. Rice (Eds.), (John Wiley and Sons, 1987).

3. B. T. Sutcliffe, Fundamentals of computational chemistry, in *Computational Techniques in Quantum Chemistry*, G. H. E. Diercksen, B. T. Sutcliffe and A. Veillard (Eds) (Reidel, Boston, 1975).

4. J. Delhalle, M. H. Delvaux, J. G. Fripiat, J. M. Andre and J. L. Calais, *J. Chem. Phys.* **88**, 3141(1988).

5. J. Delhalle, J. Cizek, I. Flamant, J. L. Calais and J. G. Fripiat, *J. Chem. Phys.* **101**, 10717 (1994).

6. C. Pisani, R. Dovesi and C. Roetti, *Hartree-Fock ab initio Treatment of Crystalline Systems, Vol 48 of Lecture notes in Chemistry* (Springer Verlag, Berlin, 1988).

7. P. Durand and J.-P. Malrieu, *Ab initio Methods in Quantum Chemistry*, I. Prigogine and S. A. Rice (Eds), (John Wiley and Sons, 1987) pp. 321-412.

8. A. K. Chakraborty, *Polymer and Solid Interfaces* (IOP publishing, 1992), pp 1–35.

9. G. Nicolas and P. Durand, *J. Chem. Phys.* **72**, 453 (1980).

10. J. A. Pople and D. L. Beveridge, *Approximate Molecular Orbital Theory.* (McGraw-Hill, 1970).

11. R. Sustmann, J. E. Williams, M. J. S. Dewar, L. C. Alien and P. R. Schleyer, *J. Am. Chem. Soc.* **91**, 5350 (1969).

12. R. C. Bingham, M. J. S. Dewar and D. H. Lo, *J. Am. Chem. Soc.* **97**, 1285(1975).

13. M. J. S. Dewar and W. Thiel, *J. Am. Chem. Soc.* **99**, 4899(1977).

14. M. J. S. Dewar, E. G. Zoebisch, E. E Healy and J. J. P. Stewart, *J. Am. Chem. Soc.* **107**, 3902(1985).

15. J. J. P. Stewart, *J. Comp. Chem.* **10**, 221(1989); J. J. P. Stewart, *Q. C. P. E.* no. 581.

16. J.-M. Andre, L. A. Burke, J. Delhalle, G. Nicolas and P. Durand, *Int. J. Quantum. Chem. Symp.* **13**, 283 (1979),

17. P. Hohenberg and W. Kohn, *Phys. Rev.* **136**, B864 (1964).

18. W. Kohn and L. J. Sham, *Phys. Rev.* **104**, A1133 (1965).

19. U. v. Barth and L. Hedin, J. *Phys. C: Solid State Phys.* **5**, 2064 (1972); S. H. Vosko, L. Wilk and M. Nusair, *Can. J. Phys.* **58**, 1200 (1980).

20. A. D. Becke, *Phys Rev. A* **38**, 3098 (1988).

21. R. O. Jones in *Ab initio Methods in Quantum Chemistry*, I. Prigogine and S. A. Rice (Eds), (John Wiley and Sons, 1987).

22. UniChem, *Cray Research, Inc.*.

23. Gaussian 92, Gaussian 94, *Gaussian, Inc.*

24. R. D. Amos and J. E. Rice, CADPAC: *The Cambridge Analytic Derivatives Package.*

25. H. B. Schlegel in *Ab initio Methods in Quantum Chemistry*, I. Prigogine and S. A. Rice (Eds), (John Wiley and Sons, 1987).

26. R. S. Mulliken, *J. Chem Phys.* **23**, 1833; 1841; 2338; 2343.

27. A. E. Reed, R. B. Weinstock and E Weinhold, *J. Chem. Phys.* **83**, 735 (1995).

28. S. Stafström, J. L. Bredas, M. Lögdlund, and W. R. Salaneck, *J. Chem. Phys.* **99**, 7938 (1993).

29. T. Koopmans, *Physica (Utrecht)* **1**, 104 (1933).

30. P. Bagus, *Phys. Rev.* **139**, A619 (1965).

31. H. J. Jensen, P. Jorgensen and H. Ågren, *J. Chem. Phys.* **87**, 451 (1987).

32. M. Boman, H. Ågren and S. Stafström, *J. Phys. Chem.*, in press.

Chapter 3

Experimental methods

3.1 Survey of measurement methods

Here we survey a series of possible surface-sensitive measurements
which in principle can be used to study the surfaces of conjugated
polymers and the early stages of metal interface formation. We then
motivate the use of photoelectron spectroscopy.

NEXAFS, or near edge X-ray absorption fine-structure spectroscopy,
is one method of particular interest, which provides both chemical
and electronic information[1]. NEXAFS, as photoelectron spectroscopy,
is a photon-in-electron-out spectroscopy, and, therefore, surface
sensitive. The basis for NEXAFS is carried out using polarized,
monochromatized synchrotron radiation, which leads to sensitivity
to molecular orientation, which can be of particular use in studying
ordered polymer systems[2,3]. Also, two electron probe methods
should be mentioned. The first is high resolution low energy electron
energy loss spectroscopy, HREELS[4]. Generally somewhat destructive
to organic surfaces, and fraught with serious electronic charging
problems when used on electrically insulating organic surfaces, some
success has been achieved with HREELS in special cases. Both
electronic structure and vibrational information have been obtained
by Pireaux and co-workers[5]. Polarization effects are inherent, which
can be important in studying ordered systems. The second method is
that of high energy electron energy loss spectroscopy, which

measures the momentum-dependent energy loss function[6]. Information on the electronic transitions, as in optical spectroscopy, is obtained as a function of the momentum transfer within the electron scattering event. Free standing films of about 1000 Å thickness (or less) are required to minimize multiple scattering in transmission measurements. In addition, the equipment is very specialized and not available commercially.

Both static and dynamic SIMS, or secondary ion mass spectroscopy, are available, primarily for chemical group analysis, but these are microscopically destructive in their nature[7,8].

Optical probes of surface and interface phenomena are, of course, available, but, with a few exceptions[9], do not lend themselves to studies of surfaces under sample/device fabrication, and are inherently less surface sensitive. Nonlinear effects probed using total internal reflection are attractive[10], but generally require special sample preparation and measurement configurations. On the other hand, infrared spectroscopy of the surface and adsorbate vibrational structure is carried out using a technique named infrared reflection absorption spectroscopy, or IRAS[11]. In Fourier transform type spectrometers, FT-IRAS has an extremely high surface sensitivity. Information is not electronic, however, but indicative of the chemical nature of the surface or interface. In order to realize the full wavelength region potential (to include the 200–500 cm^{-1} region) of FTIR, however, synchrotron light sources are employed[12].

In the category of imaging studies, atomic force microscopy[13] might be proven to be most useful. To date, however, it has been difficult to obtain atomic resolution images in organic polymers, given the softness and flexibility of the samples. Usually, the measurement process results in a movement of the polymer chains. Scanning tunnelling microscopy[14], although used to image organic molecules with atomic resolution, has not been applicable to polymer surfaces and interfaces, because of the difficulties encountered in studying thick (more than about a mono-layer) organic samples. A great deal of progress has been made, however, imaging model molecules for polymers, even in solution[15].

There are many other techniques that could be mentioned, none of which have proven to be general laboratory tools, but each of which is useful in specific circumstances: in particular, Rutherford backscattering[16,17], low energy electron diffraction[18], positron diffraction[18], molecular beam techniques[19], ion scattering[20], photoelectron diffraction[21], and many more. A recent review of the development and applicability of surface science techniques is available[22]. Few of these techniques are applicable, however, to the surfaces of (thick) organic materials, because of inherent sample surface (electronic) charging effects, which degrade or completely obliterate spectra[23].

3.2 Photoelectron spectroscopy

3.2.1 Motivation

To date, the single most useful experimental method to study both the chemical and electronic structure of conjugated polymers and interfaces with conjugated polymers[24] has been photoelectron spectroscopy (collectively denoted as PES), including both traditional X-ray photoelectron spectroscopy (XPS, sometimes known to chemists as ESCA[25]) and ultraviolet photoelectron spectroscopy (UPS). When carried out with variable photon energies from synchrotron radiation, where the distinction between XPS and UPS becomes diffuse, the acronym PES is often used. The reasons for the applicability of this generalized technique (especially to materials under consideration here) are (i) there is a maximum amount of both chemical and electronic information within a single measurement method, (ii) it is essentially non-destructive to organic systems, and (iii) it is extremely surface sensitive. On the other hand, photoelectron spectroscopy is a relatively 'low energy resolution' spectroscopy compared with, for example, infrared spectroscopy. In addition, PES must be carried out in vacuum. In order to measure the kinetic energy of the photoemitted electrons, without significant collisions with molecules in the background pressure of the spectrometer, typically $p < 10^{-5}$ Torr is required. In order that surfaces remain contamination free, especially when working with, e.g., reactive metal (donor) doping of conjugated polymers, ultra-high vacuum ($p < 10^{-9}$ Torr) is required. Experimental hardware details, however, will not be discussed in this work. Below, some especially pertinent features of photoelectron spectroscopy will be pointed out, which are of interest just in the area of conjugated polymer surfaces and interfaces[7,23,24,26].

3.2.2   Brief overview of PES

The samples studied in connection with work discussed here consist of ultra-thin films of molecular solids, either polymers or condensed molecular solids. It is useful, however, to describe first the molecular photoelectron emission process, at least from a phenomenological view point, and then present the extra issues in dealing with solids.

Consider a photon of energy $h\upsilon$ absorbed by an isolated molecule, and an electron of kinetic energy E_K emitted ('photoemitted') from the molecule. The physics of the process may be represented symbolically by

$$M_o + h\upsilon \rightarrow M_+{}^* + e^-.$$

The total energy of the 'system' on the left side of the arrow (the total energy of the neutral molecule in the ground state, M_o, plus that of the photon, $h\upsilon$) is conserved as the total energy on the right side of the arrow (the excited molecular cation, $M_+{}^*$, and the kinetic energy of the escaping photoemitted electron). Unfortunately, in the types of large organic molecules considered here, the electronic structure of M_o and $M_+{}^*$ must be modelled in terms of single-electron states, with as much physics built in as possible. The theoretical issues involved, the models used, and the parameters which are input into the models are discussed in chapter 2. Several highly schematic single-electron molecular configurations useful in the present context are shown in Fig. 3.1. The panels illustrate the neutral molecule in the ground state, M_o, where all of the electrons in the molecule occupy only the lowest allowed energy levels (V_i), while the V_i^* levels are empty; the generalized excited state of the (photoionized) molecular action, $M_+{}^*$, where an electron has been removed form the core level, C_i, but the remaining electrons do not occupy the lowest possible single-electron states (the special case of ionization from the HOMO level is illustrated as M_+); and the optically excited neutral molecule, $M_o{}^*$, (where one electron has been excited from the HOMO to the LUMO in the illustration). In addition, a case of particular interest in the examples in chapter 7, is that of $M_+{}^{**}$, representing the simultaneous photoionization of C_i and an electronic excitation (HOMO to LUMO in the illustration). The case of M_+^{**} illustrates the phenomenon of 'shake-up', where an electron is excited across the HOMO–LUMO gap simultaneously with a core-electron ionization event[25,27,28].

Neglecting experimental parameters which come in to the relationship in connection with an actual measurement, the effective binding energy of the electrons in a molecule is defined 'schematically' (see below) as $E_B = h\upsilon - E_K$. Usually, both the kinetic energy and the binding energy are presented as positive numbers (sometimes, however, the binding energy is found written as a negative number); 'Let the reader beware'. With soft X-ray photons (XPS), for example, 1254.6 eV $Mg(K_\alpha)$ radiation from the most common laboratory source, both the atomic core-electron energy levels (C_i) and the

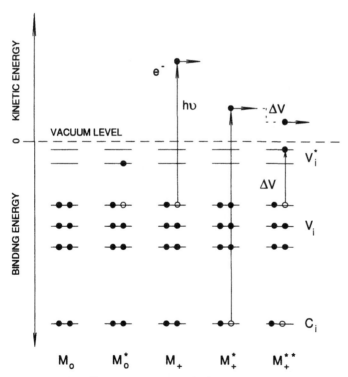

Fig. 3.1   *Illustration of the single-electron energy levels of the various states of the molecule, as discussed in the text.*

valence electron energy levels (V_i) may be studied. With ultraviolet radiation (UPS), for example 21.2 eV or 40.8 eV photons from a helium discharge source, only the valence electronic states may be studied, but with certain resolution and photoionization cross section advantages over XPS. A most central point is that, for polymer materials, *there is a one-to-one correspondence between the maxima (peaks) in the photoemitted electron kinetic energy distribution (a spectrum) and the electron energy states in the molecule (sample)*, as illustrated in Fig. 3.2. In the figure, the discrete kinetic energy values ($E_K = h\upsilon - E_B$) in the distribution of photoemitted electrons, the 'spectrum', have been broadened in order to appear to simulate experimental data.

Before discussion of the PES of polymers and other molecular solids, however, many-electron effects inherent in the spectroscopy

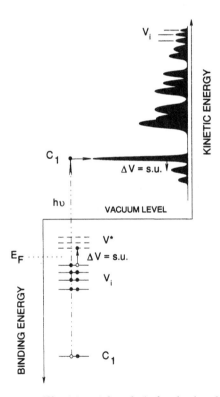

Fig. 3.2   *A hypothetical molecular electron energy level scheme is show with increasing binding energies downward. The corresponding photoelectron spectrum is simulated above, with increasing electron kinetic energies upward.*

of *isolated molecules* must be pointed out. For the present discussions, there are two relevant (and related) manifestations of the many-body nature of the photoionization process: (1) apparent *energy shifts* in recorded spectra, arising from two separate sources; and (2) satellite structure, observed on the high binding energy side of the main photoelectron line in core-electron spectra, which are a measure of the electronic structure of the molecule.

First, satellite structure on the high binding energy side of, for example, an XPS core-level line (or 'peak') corresponds to so-called 'shake-up' (referred to below as 's.u.') and 'shake-off'[25,29] effects, the former of which is illustrated, by M_+^{**}, in Fig. 3.1. Shake-off is just shake-up to the continuum rather than to an unoccupied molecular state. Considerations of (1) are important in comparisons with the results of model calculations; while (2) is of use as an indication of the electronic transitions in the molecules under study, an example of which is found in studies of the early stages of interface formation, i.e., the interactions of reactive metal atoms with conjugated polymer surfaces. Since use will be made of these effects in subsequent chapters, they are outlined briefly below.

The conventional binding energy relationship, $E_B = h\upsilon - E_K$ (the so-called Koopman's theorem[30]), is only a conceptually useful approximation. Many-electron effects lead to *intra*-molecular electronic relaxation effects not included in the simple equation[28,31]. These *intra*-molecular electronic relaxation effects are quantum in nature, but may be rationalized for simplicity in terms of a classical picture. Simultaneous with the generation of, for example, a core hole, the electrons remaining in the molecule completely screen the generated hole state[32,33], on a time scale faster than that of the escaping photogenerated electron. The nuclei in the 'lattice' (molecule), on the other hand, are too massive to respond on the same time scale (electronic relaxation is said to be fast, since the mass of the electron is small compared with that of the nuclei). The measured kinetic energy of the photoemitted electron (the main line) represents the binding energy of the electronic state from which it originated, but with a hole in it, and in the presence of the nuclei at the lattice positions of the neutral molecule in the ground state (an unusual beast).

Quantum mechanically, the photoionization event is well described within the context of the sudden approximation[34]. The main photoelectron line ($E_K = h\upsilon - E_B$), however, does not appear at the binding energy of the electronic state in the neutral molecule in the ground state (E_B). It appears 'shifted' towards apparently lower binding energy due to *intra*-molecular relaxation effects. In response to the sudden generation of the core hole, the electrons, originally in eigenstates of the neutral molecule in the ground state, are 'suddenly' frozen in the single-particle eigenstates (as well as particle-hole states and hole-continuum states) of the molecular cation (with nuclei still at the positions of the neutral molecule in the ground state), with each eigenstate carrying some intensity[35]. This is 'how' (in the classical picture of shake-up, or s.u.) the remaining electrons in the molecule screen the photogenerated hole.

The XPS spectrum observed for photoionization of a given molecular electron energy eigenstate, for example the C_i in the figures, consists of a multitude of satellite lines, each corresponding to a slightly different state of the molecular cation[27,28,31,36]. In the full spectrum, the apparent 'main line' (almost always the largest XPS peak) carries most of the intensity, but does not appear at the E_B expected from, for example, some *a priori* knowledge of the electronic structure of the neutral molecule in the ground state. Within the Hartree–Fock limit, it is rigorously true that the first-moment (M_1) of the total core-electron spectral distribution is equal in energy to the Hartree–Fock eigenvalue[36], as shown schematically in Fig. 3.3. This model is generally accepted as containing the essence of the physics of the process[27]. It has been shown in carefully chosen gaseous systems, where the full XPS spectra could be measured, that when the weighted average over all of the shake-up and shake-off spectral features is included, a very good indication of the energy of the state of the electron in the neutral system in the ground state is obtained[27]. Unfortunately, because of experimental difficulties with the inelastic electron background in spectra of solids, the full weighted spectrum is almost never obtained. The shake-up structure is still useful, however, as a 'finger printing method', as will be shown below.

Although these shake-up, or s.u., electronic transitions are indicative of the electronic structure of the molecular cation, the spectrum of s.u. structure often, but not always, resembles the optical absorption spectrum of the neutral molecule initially in the ground state. The s.u. spectra are useful, however, in studying the molecular electronic transitions at surfaces, because, as in PES itself, the spectra originate from the very surface region. In particular, s.u. spectra represent a very sensitive measure of electronic structure changes in the near-surface region of a molecular solid induced by, for example, interactions with metal atoms at the early stages of interface formation at the surfaces of conjugated polymers. Interpretation with the help of proper calculations is uncommon, however, because of the computational difficulties involved[37]. Through comparison with reference spectra (finger printing), however, s.u. spectra can and do provide much information, as will be discussed in the examples.

The next issue in the overview of PES involves surfaces of polymer thin films and condensed molecular solids. There are several issues to be mentioned: (1) *inter*-molecular polarization (relaxation) effects and (2) surface sensitivity. First, under (1), the concept of *intra*-molecular was described above. When molecules

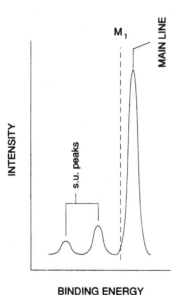

Fig. 3.3   *The relationship of the first moment of the photoelectron spectrum of a core state to the main peak and the shake-up satellite structure is illustrated*[27,36].

reside in a molecular solid, in addition to the *intra*-molecular polarization (relaxation) effects (which the remaining electrons in the molecule experience upon the photogeneration of a hole), there exist also *inter*-molecular polarization effects, arising from the polarization of the electrons in surrounding (mostly nearest neighbour) molecules in a molecular solid[38,39]. When measured relative to the same reference energy, the vacuum level, the kinetic energy of a photogenerated electron from a given molecule in the solid state will be higher than that for the same molecule in the gas phase by just the inter-molecular relaxation (polarization) energy, E_P. Various studies have shown that a value of E_P typical of hydrocarbon molecules is[26,38,39] $\backsim 1\frac{1}{2}$ eV. Inter-pmolecular relaxation is important in comparing the results of calculations of the density-of-valence-states (DOVS) with experimental UPS spectra of the same quantity. The adjustment of the energy axis of the calculated results by E_P was discussed in chapter 2.

In the UPS or XPS of solid films, the front of the sample (the polymer or condensed molecular solid film) can be maintained in electrical contact (equilibrium) with the metallic substrate, because the films employed are so thin that electron tunneling from the substrate, or numerous other effects, prevent any positive surface electronic charge from building up during the course of the measurements. In this way, electronic sample charging effects[23,40,41] are avoided in works reported herein, and thus will not be covered.

The surface sensitivity of PES arises from the short elastic mean free path for low kinetic energy electrons in solids; which can be quantified approximately as follows. The photons used usually can penetrate the sample deeper than the region near the surface from which the photogenerated electrons can escape (without inelastic scattering). The elastic mean free path, λ, is defined by the attenuation relation; the number of electrons transversing a distance d without inelastic energy loss is given by $n(d) = n(0)\exp(-\lambda/d)$, where λ is kinetic energy dependent. Above about 100 eV (electron kinetic energy), λ varies approximately[25] as $(E_K)^{\frac{1}{2}}$. Below about 100 eV, however, as $E_K \rightarrow 0$ the curve rises rapidly for metals (approaching the mean inelastic scattering length of electrons at the Fermi energy of metals). For organic molecular solids, however, λ

approaches about 5 Å (a typical inter-molecular hopping distance)[38,40]. At higher values of E_K, at several well known points, λ has been measured carefully for hydrocarbon polymers. For example, for the emission of electrons from the $\mathrm{Au}(4f_{7/2})$ level of a sample substrate, using $\mathrm{Mg}(K_\alpha)$ photons ($E_K = 1.17\,\mathrm{keV}$), through carefully measured thicknesses of polymeric hydrocarbons[42], $\lambda \approx 22\,\text{Å}$. Using the value of 22 Å, the thicknesses of thin ($\sim 100\,\text{Å}$ or less) films of condensed molecular solids on gold substrates may be estimated using the attenuation of the XPS signal from the $\mathrm{Au}(4f_{7/2})$ level, provided that appropriate precautions are taken to assure sample thickness uniformity. As a result of the energy dependence of λ, the ideal thickness of highly insulating samples, polymers or condensed molecular solids, is about 100 Å for XPS and about 50 Å for UPS. Nominally, the depth to which unique atoms can be observed is taken to be approximately 2λ. At the thicknesses listed, any signal from the polymer substrate interface is so weak that it does not interfere with studies of the front surface, and subsequent interface formation. On the other hand, in many conjugated polymers, even free-standing films of up to about 2000 Å have been studied without detrimental surface charging effects[43], presumably because of inherent electrical conductivity and/or weak photoconductivity, sufficient to prevent the build-up of surface charge.

A procedure related to the short elastic mean free path of the escaping electrons, which is useful at least in XPS, allows a measure of the surface depth distribution of elements with unique XPS signals to be obtained[44]. By varying the angle at which the escaping photogenerated electrons are collected, electrons originating from somewhat different distances from the surface are observed. The measurement is denoted as $\mathrm{XPS}(\theta)$. As a model example, consider Fig. 3.4. For normal exit, electrons emanating from unique atoms at the 'bottom' of the ordered (in the direction normal to the surface) molecular layer must traverse the molecular layer on the way to the collector (electron energy analyser); the signal at the surface is proportional to $\exp(-\lambda/d)$, where d is the distance to the surface of the 'bottom' (atoms) of the molecule. Relative to the electrons generated from the 'top' of the molecule, there is a certain relative signal obtained: $\vartheta(0) = I_B(\theta = 0)/I_T(\theta = 0)$. On the other hand, for a large exit angle, the electrons emanating from the 'bottom' of the molecule (but not those from the top) must traverse a longer distance to leave the surface of the sample. The effective distance becomes $d/\cos(\theta)$, and the signal becomes proportional to $\exp(-\lambda\cos(\theta)/d)$. Thus the ratio ϑ becomes $\vartheta(\theta) = I_B(\theta)/I_T(\theta)$, which, will be less than $\vartheta(0)$ as $\theta \to 90°$. In general, by modeling the θ-dependence of ϑ, a certain (but not really very accurate) measure of the depth distribution of a unique atom in the near surface ($\backsim 50 \text{ Å}$) region of a sample may be obtained. The measure is approximate, and the foundations were laid as long as two decades ago for metals[44] and for polymers[24,45]. In studies outlined in chapter 7, use is made of XPS(θ) only to gain an indication of whether or not metal atoms, at the early stages of interface formation on conjugated polymer surfaces, are confined to the near-surface region or not. In general, except in only a few specific cases, where the thickness of the molecular films under study were specifically chosen in order to enable an indication of what is going on at the back electrode (substrate), no depth distribution profiling has been attempted.

There are photoionization cross section effects to be considered. XPS ($\backsim 1 \text{ keV}$ soft X-ray photon excitation) is used to study the core-electron energy levels, while UPS ($\backsim 20$–40 eV photons) is used to study the valence electron energy distribution. UPS has the

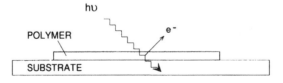

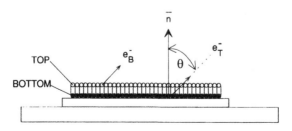

Fig. 3.4 *The principle of XPS(θ) is illustrated.*

advantage that the usual in-house photon sources have high (relative to all else in the spectroscopy) energy resolution, but the natural line widths in condensed molecular solids, from both homogeneous and inhomogeneous broadening effects, can approach 1 eV at room temperature[39,46], negating to a large extent the photon energy resolution advantage. However, there are photoionization cross section advantages with UPS, for just those purposes concerned in this book. For C(2p)-derived molecular states, the photoionization cross section is about an order of magnitude greater than that for C(2s)-derived states; rendering the most sensitivity at the π-band edge in conjugated (as well as other) polymers, just where the major interest lies. In XPS, at typical soft X-ray energies, the photoionization cross sections for C(2p)-derived states is about an order of magnitude less than that for C(2s)-derived states[47]. Thus, not only is the energy resolution somewhat better in UPS, the UPS spectra are dominated by C(2p)-derived states, while in XPS the spectra are dominated by C(2s)-derived states[48].

Finally, the width of the total UPS spectrum may be used to obtain a reasonable estimate of the work function of the solid sample under consideration[49]. The energy of the intensity cut-off of the secondary electron background in the UPS spectrum is

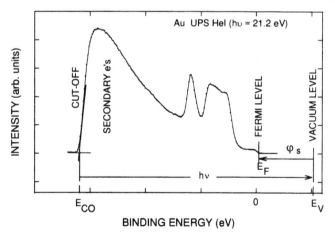

Fig. 3.5 *The determination of the work function, φ_s, from a UPS spectrum of gold.*

sensitive to the work function of the sample surface. In order to carry out a measurement which extends down to essentially zero kinetic energy electrons, the sample (substrate) is usually biased, negative relative to the entrance slits of the electron energy analyser, with anywhere from a few up to about 50 volts. The bias voltage adds to the kinetic energy of the photoemitted electrons, relative to the analyser. It easily can be shown[48,49] that on a plot of the UPS spectrum, adding the photon energy to the secondary electron cut-off energy locates the position of the vacuum level. Measuring down from the vacuum level energy to the known position of the Fermi energy (of the substrate), determines the work function, as illustrated in Fig. 3.5. Since, in equilibrium, the position of the Fermi energy is constant throughout the thickness of the ultra-thin films under discussion in this book, the position of the Fermi energy on the UPS spectrum is equal in analyser voltage to the location of the Fermi level in the metallic substrate, as observed prior to the deposition of the organic overlayer. This positioning of the Fermi level, coincident with that in the substrate, has been observed in all studies reviewed in this work, as well as in other molecular overlayer systems, on both metals and semiconductor substrates[50-52].

3.3   References

1. J. Stöhr, *NEXAFS Spectroscopy* (Springer, Berlin, 1992).
2. T. Ohta, K. Seki, T. Yokoyama, I. Morisada and K. Edamatsu, *Physica Scripta* **41**, 150 (1990).
3. K. Seki, T. Yokoyama, T. Ohta, H. Nakahaka and K. Fukuda, *Mol. Cryst. Liq. Cryst.* **218**, 85 (1992).
4. H. Ibach, *Surf. Sci.* **299/300**, 116 (1994).
5. J. J. Pireaux, in *Surface Characterization of Advanced Polymers*, L. Sabbatini and P. G. Zambonine (Eds) (VCH, Manheim, 1993).
6. J. Fink, in *Topics in Applied Physics*, J. Fuggle and J. E. Ingelsfield (Eds) (Springer, Berlin, 1992).
7. D. Briggs and M. P. Seah (Eds), *Practical Surface Analysis* (John Wiley & Sons, Chichester, 1983).
8. A. Benninghoven, *Surf. Sci.* **299/300**, 246 (1994).

9. G. Chiarotti, *Surf. Sci.* **299/300**,   541 (1994).

10. Y. R. Shen, *Surf. Sci.* **299/300**,   551 (1994).

11. Y. J. Chibal, *Surf. Sci. Rep.* **8**, 211 (1988).

12. C. J. Hirschmugl, G. P. Williams, F. M. Hoffmann and Y. J. Chabal, *Phys. Rev. Lett.* **65**, 480 (1990).

13. H. Rohrer, *Ultramicroscopy* **42-44**, 1 (1992).

14. G. Binning and H. Roher, *Rev. Mod. Phys.* **59**, 615 (1987).

15. J. P. Rabe, *Ultramicroscopy* **42-44**, 41 (1992).

16. W. K. Chu, J. W. Mayer and M. A. Nicolet, *Backscattering Spectrometry* (Academic Press, New York, 1978).

17. J. R. Bird and J. S. Williams, *Ion Beams for Materials Analysis* (Academic Press, New York, 1989).

18. C. B. Duke, in *Surface Science: The First Thirty Years*, C. B. Duke (Ed.) (North-Holland, Amsterdam, 1994), p. 24.

19. T. Engel and K. H. Rieder, *Springer Series in Modern Physics, Vol. 91* (Springer, New York, 1982).

20. J. W. Rabalais, *Science* **250**, 521 (1990).

21. C. S. Fadley, in *Synchrotron Radiation Resarch: Advances in Surface and Interface Science, Vol. 1*, R. Z. Bachrach (Ed.) (Plenum, New York, 1992).

22. C. B. Duke (Ed.), *Surface Science: The First Thirty Years* (North-Holland, Amsterdam, 1994).

23. W. R. Salaneck, *Rep. Prog. Phys.* **54**, 1215 (1991).

24. D. T. Clark and W. J. Feast (Eds), *Polymer Surfaces* (John Wiley & Sons, Chichester, 1978).

25. K. Siegbahn, C. Nordling, G. Johansson, J. Hedman, P. F. Heden, V. Hamrin, U. Gelius, T. Bergmark, L. O. Werme, R. Manne and Y. Baer, *ESCA Applied to Free Molecules* (North-Holland, Amsterdam, 1971).

26. K. Seki, in *Optical Techniques to Characterise Polymer Systems*, H. Bässler (Ed.) (Elsevier, Amsterdam, 1989), p. 1.

27. H. J. Freund, E. W. Plummer, R. W. Salaneck and R. W. Bigelow, *J. Chem. Phys.* **75**, 4275 (1981).

28. H. J. Freund and R. W. Bigelow, *Physica Scripta* **T17**, 50 (1987).

29. U. Gelius, *Physica Scripta* **9**, 133 (1974).

30. T. Koopmans, *Physica* **1**, 104 (1934).

31. G. Wendin, *Photoelectron Spectra* (Springer Verlag, Berlin, 1981).

32. J. J. Pireaux, S. Svensson, E. Basilier, P. Malmquist, U. Gelius, R. Caudano and K. Siegbahn, *Phys. Rev.* **A14**, 2133 (1976).

33. A. Naves de Brito, S. Svensson, M. P. Keane, H. Ågren and N. Correia, *Europhys. Lett.* **20**, 205 (1992).

34. L. I. Schiff, *Quantum Mechanics* (McGraw Hill, New York, 1968).

35. P. Nozieres, *Theory of Interacting Fermi Systems* (Benjamin, New York, 1964).

36. R. Manne and T. Åberg, *Chem. Phys. Lett.* **7**, 282 (1962).

37. B. Sjögren, W. R. Salaneck and S. Stafström, *J. Chem. Phys.* **97**, 137 (1992).

38. W. R. Salaneck, *Phys. Rev. Lett.* **40**, 60 (1978).

39. C. B. Duke, W. R. Salaneck, T. J. Fabish, J. J. Ritsko, H. R. Thomas and A. Paton, *Phys. Rev.* **B18**, 5717 (1978).

40. P. Nielsen, D. J. Sandman and A. J. Epstein, *Solid State Commun.* **17**, 1067 (1975).

41. W. R. Salaneck and R. Zallen, *Solid State Commun.* **20**, 793 (1976).

42. D. T. Clark and H. R. Thomas, *J. Polym. Sci., Polym. Chem. Ed.* **15**, 2843 (1977).

43. J. Rasmusson, S. Stafström, M. Lögdlund, W. R. Salaneck, U. Karlsson, D. B. Swansson, A. G. MacDiarmid and G. A. Arbuckle, *Synth. Met.* **41-43**, 1365 (1991).

44. C. S. Fadley, R. J. Baird, W. Siekhaus, T. Novakov and L. Bergstrom, *J. Elec. Spec.* **4**, 93 (1974).

45. W. R. Salaneck, H. W. Gibson, F. C. Bailey, J. M. Pochan and H. R. Thomas, *J. Polym. Sci., Polym. Lett. Ed.* **16**, 447 (1978).

46. W. R. Salaneck, C. B. Duke, W. Eberhardt and E. W. Plummer, *Phys. Rev. Lett.* **45**, 280 (1980).

47. K. Siegbahn, *J. Elec. Spec.* **5**, 3 (1974).

48. W. D. Grobman and E. E. Koch, in *Photoemission From Solids*, M. Cardona and L. Ley (Eds) (Springer-Verlag, Berlin, 1979).

49. E. W. Plummer and W. Eberhardt, *Adv. Chem. Phys.* **49**, 533 (1982).

50. T. R. Ohno, Y. Chen, S. E. Harvey, G. H. Kroll, J. H. Weaver, R. E. Haufler and R. E. Smalley, *Phys. Rev.* **B44**, 13747 (1991).

51. G. Gensterblum, K. Hevesi, B. Y. Han, L. M. Yu, J. J. Pireaux, P. A. Thiry, D. Bernaerts, S. Amelinckx, G. Van Tendeloo, G.

Bendele, T. Buslaps, R. L. Johnson, M. Foss, R. Feidenhans'l and G. Le Lay, *Phys. Rev.* **B50**, 11981 (1994).

52. A. J. Maxwell, P. A. Brühwiler, A. Nilsson and N. Mårtensson, *Phys. Rev.* **B49**, (15), 10717 (1994).

Chapter 4

Materials

4.1 Molecular solids

Molecular solids are composed of discrete units which retain their identities as molecules in the solid phase, for example, when condensed from the gas phase into thin films. Molecular solids may be single crystals, polycrystalline or amorphous in structure[1]. Bonding within the molecules is usually covalent, and is localized to within the molecular units. There are usually no *inter*-molecular covalent bonds, only weak Van der Waals forces[2]. On the other hand, there are systems which are molecular yet more firmly bound in the solid, e.g., the amino acids and other hydrogen-bonded molecular solids. Hydrogen bonding in the zwitterionic state of condensed molecular solids has been studied by photoelectron spectroscopy, the primary experimental tool used in the results presented in this book[3].

In chapter 7, all works discussed on model molecular systems for conjugated polymers refer to condensed molecular solid ultra-thin films, generally prepared by condensation of molecules from the effusion of a Knudsen-type cell, in UHV, on to clean metallic substrates held at low temperatures. Clean is defined as atomically clean as determined by core-electron level XPS, such that there is intimate contact between the molecules at the substrate–film interface, without the influence of, for example, a metallic oxide, hydrocarbon or other contamination. In general, non-cross-linked polymers, and linear conjugated polymers in particular, comprise a special case of molecular solids. Exceptions are pointed out, as necessary.

4.2   π-Conjugated polymers

In linear polymer molecules which are *not* conjugated, the electronic structure of the chain of atoms or chemical groups which comprises the backbone of the macromolecule consists of only σ-bands (possibly with *localized* π-electronic levels). The large electron energy gap, E_g, in the σ-system renders these polymer materials electrically insulating and transparent to visible light. *π-conjugated polymers*, on the other hand, consist of a regularly alternating system of single (C–C) and double (C=C) bonds; a condition which leads to a lower E_g, in the delocalized π-system.

The geometrical structures of several common conjugated polymers are shown in Fig. 4.1, where the necessary (but not sufficient) condition for conjugation, the alternation of the single and double bonds along the polymer backbone, can be seen. The unique electronic properties of conjugated polymers derive from the presence of π-electrons, the wavefunctions of which are delocalized over long portions of (if not the entire) polymer chain, when the molecular structure of the

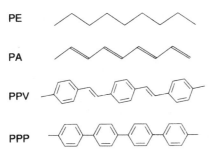

Fig. 4.1   *The chemical structures of several relevant polymers are illustrated. There is a carbon atom at each vertex, and the hydrogen atoms are not shown. PE(CH₂): polyethylene, or PE, with only the C-H single bonds shown; PA: trans-polyacetylene; PPV: poly(para-phenylenevinylene); and PPP: poly(para-phenylene). The lower three polymers are conjugated, according to the alternating single and double bond system.*

backbone is (at least approximately) planar. It is necessary, therefore, that there are not large torsion angles at the bonds, for example, joining aromatic groups of the lower two polymers in Fig. 4.1, which would decrease the delocalization of the π-electron system[4]. The essential properties of the delocalized π-electron system, which differentiate a typical conjugated polymer from a conventional polymer with σ-bands, are as follows: (i) the electronic (π-) band gap, E_g, is relatively small (~ 1–$3\frac{1}{2}$ eV), with corresponding low energy electronic excitations and semiconductor behaviour; (ii) the polymer molecules can be rather easily oxidized or reduced, usually through charge transfer with atomic or molecular dopant species, to produce *conducting polymers*; (iii) net charge carrier mobilities in the conducting state are large enough that high electrical conductivities are realized; and (iv) the charge carrying species are not free electrons or holes, but quasi-particles, which, under certain conditions, may move relatively freely through the materials[5,6].

The magnitude of the room temperature electrical conductivity in thin films of iodine-doped polyacetylene, ordered by stretch-alignment (to relatively parallel polymer chains), is about 4×10^5 S/cm, i.e., almost as large as that of a single crystal of copper[7]. Electrical conductivity as high as 3×10^4 S/cm is reached in doped *fibre* materials[8]; this is obtained through the use of highly ordered polyacetylene fibres fabricated by tensile drawing. In these fibrous materials, true macroscopic delocalization of the one-dimensional π-electron system is observed. Also, high electrical conductivities, ranging between about 10^3 and 10^4 S/cm, are found in stereoregular polythiophene[9] and highly stretched polyaniline[10]. No true metallic conductivity has been measured yet (for example, electrical conductivity does not increase all the way down to low temperatures)[11]. It appears that the 'best' current polyacetylenes and polyanilines are on the border line of the metal–insulator transition. Further improvements in materials processing (reduction of disorder) could lead to the development of true metallic systems.

The materials properties of conjugated polymers have improved significantly over the last five years, especially because of major developments in controlled synthesis and processing technology[12,13]. Recent overviews of the general state-of-the-art can be found in the

Proceedings of the 81st Nobel Symposium on Conducting Polymers[14], the book of Brédas and Silbey[4], as well as, for example, a recent collection of overview articles[15].

The intrinsic low-dimensional geometrical nature of linear polymer chains, as well as the general property of conjugated organic molecules that the geometric structure is dependent upon the ionic state of the molecule (strong electron(hole)–lattice interactions), leads to the existence of unusual charge carrying species. These species manifest themselves, through either optical absorption in the neutral system or charge transfer doping, as self-localized electronic states with energy levels within the otherwise forbidden electron energy gap[5,16–19]. For example, in n-type doping, although the very first single electron on the (any) isolated *trans*-polyacetylene polymer chain goes in as a polaron[5,18], polarons readily combine to form a lower energy state, that of the soliton (after the mathematics which describes its behaviour[16,17]). Although solitons must be formed in pairs[20], usually a single specie is diagrammed for convenience. An isolated soliton, containing an excess electron, leads to the formation of a new electronic state within the energy gap. In Fig. 4.2 are shown schematic representations of the neutral soliton (top), the negatively charged soliton (middle) and the positively charged soliton (bottom). The bond alternation associated with these states has been abbreviated for convenience[5,20]. In non-degenerate ground-state polymers, however, the individual solitons interact (combine) to form spinless bipolarons, a lower energy configuration. Thus, generally, depending upon the symmetry of the ground state, the charge carrying species are charged polarons, spinless charged solitons, or spinless charged bipolarons[5,16–18,20,21]. The bipolaron state for poly(p-phenylenevinylene), or PPV, a non-degenerate conjugated polymer, is sketched in Fig. 4.3. These species, solitons, polarons and bipolarons, *represent the lowest energy eigenstates of the coupled electron (hole)–lattice systems*[22], and are responsible for the unusual electrical, magnetic and optical properties of conjugated polymers.

Conjugated polymers, in the undoped state, present remarkable nonlinear optical properties[23]. The material which has been most thoroughly investigated ever since 1976[24] is polydiacetylene *para*-toluene sulfonate (PTS). The polydiacetylene backbone consists of a succession of single–double–single–triple carbon–carbon bonds. It also is interesting to note that polydiacetylene can be obtained in the form of single crystals of very high quality (which is presently unique among the organic conjugated polymers) or crystalline thin films[25]. Recent studies by Stegeman and co-workers[26,27] demonstrate that PTS figures of merit are excellent for all-optical applications in the telecommunications window (1.3 μm; 1.55 μm). Other conjugated polymers that currently are studied for nonlinear applications include polyacetylene, polyarylenes and polyarylene vinylenes[23].

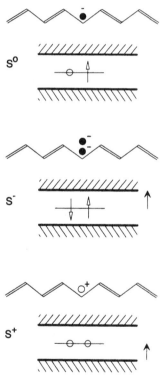

Fig. 4.2 *A short segment of* trans-*polyacetylene is shown with an abrupt (idealized) reversal of the bond alternation pattern (see text). Top: a neutral soliton with an unpaired spin and an energy state near the middle of the electron energy gap. Middle: the addition of an electron results in the formation of a spinless negatively charged soliton. Bottom: the extraction of an electron from the top results in the formation of a spinless positive soliton. The optical transitions associated with the charged solitons are indicated as arrows on the right.*

Below, the one-dimensional band structure of simple linear conjugated polymers is outlined, followed by a brief description of the use of some model π-conjugated molecules for conjugated polymers. Then, to close out the chapter, the description of the properties of conjugated polymers is extended to issues of optical absorption and photo-luminescence.

4.3   One-dimensional band structure of linear conjugated polymers

The focus of this section is the *electronic structure of a polymer molecule*, which is a reasonably accurate (to be defined) estimate of the energy eigenfunctions and corresponding eigenvalues of a molecule, defined by the identity and (average equilibrium) positions of the nuclei. When the molecule is so large that the energy eigenvalues tend to merge into apparent bands, as in a long polymer molecule, the desired electronic structure is referred to as the *electronic band structure*.

A wide variety of literature exists discussing energy bands in solids from many different points of view[28–30]. As discussed in chapter 2, for the case of stereoregular conjugated polymers, the essential feature involves extending the molecular orbital eigenvalue problem to systems with one-dimensional periodic boundary conditions, i.e., over regularly repeating monomeric units. Although even the linear macromolecule is really '3-D' over large distances, the periodicity is taken only in 1-D over short distances of gentle curvature. The

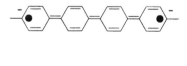

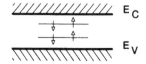

Fig. 4.3   *The bipolaron state in poly(p-phenylenevinylene) is shown schematically.*

translational symmetry of the polymer chain implies that the solutions of the Schrödinger equation must be of the special Bloch form:

$$\Psi(k, r) = u_k(r) \exp(ikx), \qquad (7)$$

where $u_k(r)$ has the period of the lattice (the unit repeating along only one-dimension, say, x), and k is the crystal momentum along the direction of x. Periodicity in reciprocal space implies that $\Psi(k, r) = \Psi(k+K, r)$, where K is the reciprocal lattice vector. The first Brillouin zone is defined by the region between $-\pi/a \le k \le \pi/a$, where a is the magnitude of the real space lattice vector. The first Bragg reflections and the first forbidden electron energy gap occur at the first Brillouin zone. The simple one-dimensional energy band structure of an electron moving in a potential with repeat distance 'a' is illustrated in Fig. 4.4. The electronic energy band structure of an arbitrary one-dimensional system will contain many overlapping bands, usually of different symmetry, and spread out over different binding energies. The N electrons of the system will occupy the lowest (deepest binding energy) bands.

The geometrical structure of *trans*-polyacetylene, which is the stable polyacetylene stereoisomer at room temperature, is illustrated in Fig. 4.1. Of particular interest is the series of alternating single and double bonds. Since the carbon atoms are sp²-hybridized (each carbon atom has three nearest neighbours), the remaining p_z atomic

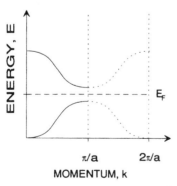

Fig. 4.4 *The energy band of an electron moving in a simple periodic potential of repeat distance 'a' is shown, both in the first Brillouin zone (π/a) and extended to the second zone (2π/a).*

orbitals, one on each carbon atom and oriented perpendicular to the planar zig zag chain, result in a so-called π-band electronic energy band. The calculated energy band structure of *trans*-polyacetylene was discussed[31] in chapter 2. In Fig. 4.5, the calculated band structure is compared with measured UPS spectra.

Normally, a band structure is presented with the energy along the vertical (*y*-axis) and the momentum vector **k** along the horizontal (*x*-axis). The band structure in the figure is presented here rotated by 90°, to enable a more direct comparison with the experimental spectrum. To facilitate the comparison between the UPS data and the calculated band structure, the density-of-valence-states, $\rho(E)$, (or DOVS) described in chapter 2, is included. The highest occupied energy band is the π-band, which can be seen clearly as the band edge in the UPS spectra. The position of the (π-) band edge, and changes in the electronic structure in the region of the edge, are of central importance in studies of conjugated polymer surfaces and the early stages of formation of the polymer–metal interface. Clearly, the UPS spectra, sensitive to only the top-most molecules of the film, are appropriate in such studies.

4.4   Molecular models for conjugated polymers

In many instances, it is an advantage, from both an experimental and theoretical point of view, to study appropriately chosen model molecules for conjugated polymers in terms of both the surface electronic structure and the early stages of metal–polymer interface formation. Results obtained may then be compared with the results of equivalent experiments on real conjugated polymers[32]. In particular, when dealing with the surfaces and interfaces of ultra-thin films, as in the present case, model molecules can be chosen for which physical vapour deposition (PVD) in UHV is possible. Among the advantages are:

- the PVD process is a self-distillation process, so that chemically very pure films may be prepared;
- homogeneous films of uniform thickness may be prepared directly in UHV;

- the film thickness can be controlled as necessary, for example, when the homogeneity of doping or the depth extent of the metal–polymer (model molecule) interface should be studied;
- spectra obtained are almost always more resolved ('cleaner') than those for a corresponding polymer thin film; and

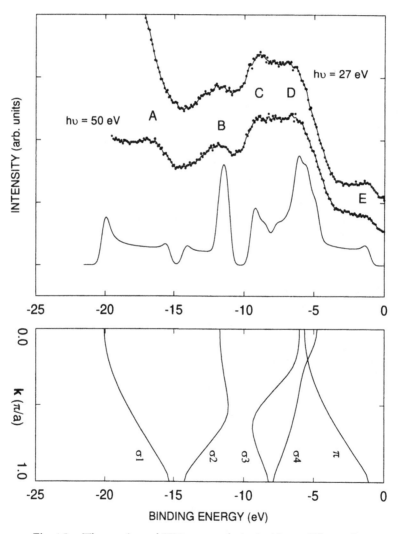

Fig. 4.5   *The experimental UPS spectra, obtained with two different photon energies, are compared with the VEH-derived band structure of* trans-*polyacetylene. An intermediate step, the DOVS calculated from the band diagram, are shown for comparison.*

■ model calculations often may be made at a higher quantum chemical level because of the finite size of the molecules involved.

Examples of the usefulness of model molecules are given in chapter 7. However, as a representative, in Fig. 4.6 is shown a segment of *trans*-polyacetylene, denoted simply as $(CH)_x$, a segment of poly(*para*-phenylenevinylene), or PPV, along with a molecule which is a model for both $(CH)_x$[33] and PPV[34]; namely α,ω-diphenyltetradecaheptaene, abbreviated as 'DP7'; indicating a diphenyl (D) polyene (P) with seven C=C double bonds (7) in the polyene portion of the molecule. The frontier electronic structure of the shorter diphenylpolyenes has received some attention previously[35,36]. For the diphenylpolyenes in general, and in DP7 in particular, the frontier molecular wavefunctions (molecular orbitals) are almost completely localized on the polyene portion of the molecule. This is the case in DP7 for both the two highest occupied (π) wavefunctions

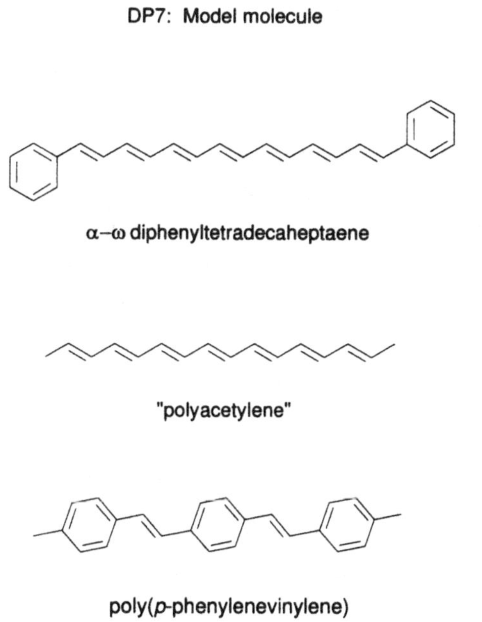

DP7: Model molecule

α–ω diphenyltetradecaheptaene

"polyacetylene"

poly(*p*-phenylenevinylene)

Fig. 4.6   *A particularly useful model molecule for studying certain issues involved with poly(p-phenylene) surfaces and interfaces, α, ω-diphenyltetradecaheptaene, denoted as 'DP7', is shown at the top. The similarities with both trans-polyacetylene and poly(p-phenylene) are obvious.*

and the lowest two unoccupied (π^*) wavefunctions[37]. The electronic structure of the HOMO is a good approximation to the upper π-band edge in $(CH)_x$. The presence of both a polyene part and the two phenyl-groups provides several similarities to PPV, for which DP7 has been extensively studied[34]. In addition, there are other diphenyl-polyenes, which have been studied somewhat less in the present context, but with the same motivation[38,39]. In a related system, a significant number of studies have been carried out on oligomers of poly(p-phenylene), or PPP, as model systems for the interactions of metal atoms with conjugated polymers[40].

Some commonly used model molecules, and their polymeric counterparts, are listed in Table 4.1. References will be cited in connection with the examples given in subsequent chapters.

A particularly interesting fact is that in the case of α-sexithiophene, or α6T, work on prototype devices such as MISFETs has shown that thin films of α6T molecules actually exhibit better performance characteristics than the corresponding polymer, polythiophene, and rival those of amorphous silicon[41]. α6T also has been used in light emitting diodes[42–44].

Studies of the early stages of formation of the aluminium-on-polythiophene interface have been interpreted with the help of model studies of aluminium atoms on condensed molecular solid films of α6T[45], as will be illustrated in chapter 7. Other studies using model molecules for conjugated polymers include the investigation of the interaction between metals and conjugated molecules, including oligomers of conjugated polymers[38,40,46–50].

Table 4.1 *Some typical model molecules for conjugated polymers*

Conjugated polymer	Model molecules
polythiophene	α-sexithiophene, other thiophene oligomers
polyacetylene	1,3,5-hexatriene, DP7, β-carotene
poly(p-phenylene)	biphenyl, longer oligomers of PPP
poly(p-phenylenevinylene)	DPx with $2 < x \le 7$

4.5 References

1. R. Zallen, *The Physics of Amorphous Solids* (Wiley-Interscience, New York, 1983).

2. C. B. Duke, W. R. Salaneck, A. Paton, K. S. Liang, N. O. Lipari and R. Zallen, in *Structure and Excitations of Amorphous Solids*, G. Lucovsky and F. L. Galeener (Eds) (AIP, New York, 1976), p. 23.

3. R. W. Bigelow and W. R. Salaneck, *Chem. Phys. Lett.* **89**, 430 (1982).

4. J. L. Brédas and R. Silbey, *Conjugated Polymers* (Kluwer, Dordrecht, 1991).

5. A. J. Heeger, S. Kivelson, J. R. Schrieffer and W. P. Su, *Rev. Mod. Phys.* **60**, 781 (1988).

6. W. R. Salaneck and J. L. Brédas, *Solid State Communications, Special Issue on 'Highlights in Condensed Matter Physics and Materials Science'* **92**, 31 (1994).

7. H. Naarmann and N. Theophilou, *Synth. Met.* **22**, 1 (1987).

8. Y. Cao, P. Smith and A. J. Heeger, *Polymer* **32**, 1210 (1991).

9. R. D. McCullough and S. P. Williams, *J. Am. Chem. Soc.* **115**, 11608 (1993).

10. A. G. MacDiarmid, in *Conjugated Polymers and Related Materials: The Interconnection of Chemical and Electronic Structure*, W. R. Salaneck, I. Lundström and B. Rånby (Eds) (Oxford University Press, Oxford, 1993), p. 73.

11. Z. H. Wang, C. Li, E. M. Scherr, A. G. MacDiarmid and A. J. Epstein, *Phys. Rev. Lett.* **66**, 1745 (1991).

12. A. J. Heeger and P. Smith, in *Conjugated Polymers*, J. L. Brédas and R. Silbey (Eds) (Kluwer, Dordrecht, 1991), p. 141.

13. C. B. Gorman and R. H. Grubbs, in *Conjugated Polymers*, J. L. Brédas and R. Silbey (Eds) (Kluwer, Dordrecht, 1991), p. 1.

14. W. R. Salaneck, I. Lundström and B. Rånby (Eds), *Conjugated Polymers and Related Materials: The Interconnection of Chemical and Electronic Structure* (Oxford University Press, Oxford, 1993).

15. J. P. Farges (Ed.), *Organic Conductors: Fundamentals and Applications* (Marcel Dekker, New York, 1994).

16. M. J. Rice, *Phys. Lett.* **A71**, 152 (1979).

17. W. P. Su, J. R. Schrieffer and A. J. Heeger, *Phys. Rev. Lett.* **42**, 1698 (1979).

18. K. Fesser, A. R. Bishop and D. K. Campbell, *Phys. Rev. B* **27**, 4804 (1983).

19. M. Schott and M. Nechtschein, in *Organic Conductors: Fundamentals and Applications,* J. P. Farges (Ed.) (Marcel Dekker, New York, 1994), p. 495.

20. J. L. Brédas, R. R. Chance and R. Silbey, *Phys. Rev. B* **26**, 5843 (1982).

21. S. A. Brazovskii and N. Kirova, *JETP Lett.* **33**, 1 (1981).

22. D. Baeriswyl, in *Electronic Properties of Conjugated Polymers III,* H. Kuzmany, M. Mehring and S. Roth (Eds) (Springer-Verlag, Berlin, 1989), p. 54.

23. J. L. Brédas, C. Adant, P. Tackx, A. Persoons and B. M. Pierce, *Chem. Rev.* **94**, 243 (1994).

24. C. Sauteret, J. P. Hermann, R. Frey, F. Pradère, J. Ducuing, R. H. Baughman and R. R. Chance, *Phys. Rev. Lett.* **36**, 956 (1976).

25. J. Berrehar, P. Hassenforder, C. Laperssone-Meyer and M. Schott, *Thin Solid Films* **190**, 181 (1990).

26. G. I. Stegeman, *Technical Draft of 4th Iketani Conference, Hawaii, May 1994* (1994).

27. B. Lawrence, W. E. Torruelas, M. Cha, M. L. Sundheimer, G. I. Stegeman, J. Meth, S. Etemad and G. L. Baker, *submitted* (1995).

28. C. Kittel, *Introduction to Solid State Physics* (John Wiley & Sons, New York, 1986).

29. R. Hoffman, *Solids and Surfaces: A Chemist's View of Bonding in Extended Structures* (VCH, Weinheim, 1988).

30. J. M. André, J. Delhalle and J. L. Brédas, *Quantum Chemistry Aided Design of Polymers* (World Scientific, Singapore, 1991).

31. S. Stafström and J. L. Brédas, *Phys. Rev. B* **38**, 4180 (1988).

32. M. Lögdlund, P. Dannetun and W. R. Salaneck, in *Handbook of Conducting Polymers,* J. Reynolds, R. L. Elsenbaumer and T. Skotheim (Eds) (Marcel Dekker, New York, 1996), in press.

33. M. Lögdlund, P. Dannetun, S. Stafström, W. R. Salaneck, M. G. Ramsey, C. W. Spangler, C. Fredriksson and J. L. Brédas, *Phys. Rev. Lett.* **70**, 970 (1993).

34. C. W. Spangler, E. G. Nickel and T. J. Hall, *Am. Chem. Soc., Div.*

Polym. Chem. **28**, 219 (1987).

35. B. S. Hudson, J. N. A. Ridyard and J. Diamond, *J. Am. Chem. Soc.* **98**, 1126 (1976).

36. K. L. Yip, N. O. Lipari, C. B. Duke, B. S. Hudson and J. Diamond, *J. Chem. Phys.* **64**, 4020 (1976).

37. M. Lögdlund, P. Dannetun, B. Sjögren, M. Boman, C. Fredriksson, S. Stafström and W. R. Salaneck, *Synth. Met.* **51**, 187 (1992).

38. L. M. Tolbert and J. A. Schomaker, *Synth. Met.* **41-43**, 169 (1991).

39. P. Dannetun, M. Lögdlund, C. Fredriksson, C. Fauquet, S. Stafström, C. W. Spangler, J. L. Brédas and W. R. Salaneck, *J. Chem. Phys.* **100**, 1065 (1994).

40. M. G. Ramsey, D. Steinmueller and F. P. Netzer, *Phys. Rev.* **B42**, 5902 (1990).

41. F. Garnier, in *Conjugated Polymers and Related Materials: The Interconnection of Chemical and Electronic Structure*, W. R. Salaneck, I. Lundström and B. Rånby (Eds) (Oxford University Press, Oxford, 1993), p. 273.

42. G. Grem, V. Martin, F. Meghdadi, C. Paar, J. Stampfl, J. Sturm, S. Tasch and G. Leising, *Synt. Met. ICSM94* (1995), in press.

43. J. Stampfl, S. Tasch, G. Leising and U. Scherf, *Synt. Met. ICSM94* (1995), in press.

44. F. Geiger, M. Stoldt, H. Schweizer, P. Bäuerle and E. Umbach, *Adv. Mater.* **5**, 922 (1993).

45. M. Lögdlund, P. Dannetun, C. Fredriksson, W.R. Salaneck, and J. L. Brédas, submitted.

46. P. Dannetun, M. Boman, S. Stafström, W. R. Salaneck, R. Lazzaroni, C. Fredriksson, J. L. Brédas, R. Zamboni and C. Taliani, *J. Chem. Phys.* **99**, 664 (1993).

47. C. Fredriksson and J. L. Brédas, *J. Chem. Phys.* **98**, 4253 (1993).

48. C. Tanaka and J. Tanaka, *Bull. Chem. Soc. Jpn.* **66**, 357 (1993).

49. M. G. Ramsey, D. Steinmüller, M. Schatzmayr, M. Kiskinova and F. P. Netzer, *Chem. Phys.* **177**, 349 (1993).

50. D. Steinmüller, M. G. Ramsey and F. P. Netzer, *Phys. Rev.* **B47**, 13323 (1993).

Chapter 5

Device motivation for interface studies

5.1 Introduction

The application we have in mind for the metal–polymer interfaces discussed in this book is primarily that where the polymer serves as the electroactive material (semiconductor) in an electronic device and the metal is the electric contact to the device. Metal–semiconductor interfaces, in general, have been the subject of intensive studies since the pioneering work of Schottky, Strömer and Waibel[1], who were the first to explain the mechanisms behind the rectifying behaviour in this type of asymmetric electric contact. Today, there still occur developments in the understanding of the basic physics of the barrier formation at the interface, and a complete understanding of all the factors that determine the height of the (Schottky) barrier is still ahead of us[2].

The aim of this chapter is to present a simple but general band structure picture of the metal–semiconductor interface and compare that with the characteristics of the metal–conjugated polymer interface. The discussion is focused on the polymer light emitting diode (LED) for which the metal–polymer contacts play a central role in the performance of the device. The metal–polymer interface also applies to other polymer electronic devices that have been fabricated, e.g., the thin-film field-effect transistor[3], but the role of the metal–polymer interface is much less cruical in this case and therefore less relevant for detailed studies of the type discussed here.

The materials (metals and conjugated polymers) that are used in LED applications were introduced in the previous chapter. The polymer is a semiconductor with a band gap of 2-3 eV. The most commonly used polymers in LEDs today are derivatives of poly(p-phenylene-vinylene) (PPV), poly(p-phenylene) (PPP), and polythiophene (PT). These polymers are soluble and therefore relatively easy to process. The most common LED device layout is a three layer component consisting of a metallic contact, typically indium tin oxide (ITO), on a glass substrate, a polymer film (∼ 1000 Å thick), and an evaporated metal contact[4]. Electric contact to an external voltage supply is made with the two metallic layers on either side of the polymer.

The principles behind electroluminescence are discussed below. In a very simplified picture, light is produced by recombination of electrons and holes. One limiting factor for the efficiency of the device is the number of electrons and holes that are injected into the polymer. Since both electrons and the holes are needed in order to produce light, the efficiency of the device is limited by the number of minority carriers. The electron-injecting contact should have a work function that matches the conduction band of the polymer. In relative terms, this implies a metal with a low work function. The hole-injecting contact should consequently match the valence band of the polymer and have a relatively high work function. The most common choice of metal for the hole contact is indium tin oxide (ITO), which, because of its low plasma frequency, is transparent in the visible and ideal for luminescence applications. The work function for ITO is nominally 4.7 eV; but varies by at least 0.2 eV depending upon the suppliers and especially upon the handling and surface treatment[5].

5.2 Metal–semiconductor interfaces

In the following discussion of the electronic properties of metal–semiconductor interfaces, the properties of the electron-injecting contact are taken as the example for contacts. Most of our studies of metals on polymers have involved low work function metals with the application as electron-injecting contact in LED devices in mind. As stated above, in the device this contact is evaporated onto the polymer. This method is very similar to the method used in preparing metal–polymer contacts in actual LEDs, which makes a comparison among these interfaces possible, in sharp contrast to the polymer interface with the hole-injecting contact. This latter interface usually is formed by spin casting a film onto the contact. The interfacial interactions are in general different if the polymer is deposited (gently) onto the metal or if individual metal atoms hit the polymer surface with high velocity. Therefore, the most relevant application of our model systems is as electron-injecting contacts.

The conjugated polymer is almost never absolutely pristine, there is often some residual doping present in the material. The Fermi energy of the polymer is therefore not an intrinsic property of the polymer but depends on the amount of dopants present. Usually, there is a small concentration of p-type dopants in the material, which shift the Fermi energy downward compared to the pristine polymer[6]. The thin layers of oligomers prepared in ultra-high vacuum which are the focus of our studies are very close to pristine. The corresponding energy diagram is therefore very close to that of a perfect semiconductor, with the Fermi energy lying exactly at midgap. A second, and also very important, difference between the interfaces that are formed in clean but ambient atmosphere and those formed under UHV conditions is that in the latter case there is no extrinsic insulating oxide layer at the interface. The interfacial chemistry is entirely due to chemical reactions between the metal and the polymer. This is not the case for the electron-injecting contact in LEDs, which is created in the presence of oxygen.

Suppose that the metal and the semiconductor are both electrically neutral and separated from each other. Since the metal at the electron-injecting contact is assumed to have a low work function, the Fermi energy of the metal lies above that of the semiconductor, close to its conduction band. If the metal and the semiconductor are connected electrically, electrons will flow from the metal to the polymer in order to establish equilibrium, which is obtained when the Fermi energies of the two materials are aligned. Because of this flow of charge, the materials are no longer neutral after electric

contact is established. Naturally, the amount of charge transfer
between the metal and the semiconductor depends on the initial
difference in Fermi energies of the interface materials. Given the
differences in the Fermi energies shown in Fig. 5.1a, electric contact
results in the situation where positive charge resides on the surface of
the metal (typically with a depth of the Thomas–Fermi screening
distance ᵔ 0.5 Å). The corresponding negative charge gets a certain
distribution in the semiconductor layer. If the semiconductor is of
n-type (see Fig. 5.1a), the electronic charge from the metal goes into
the conduction band of the semiconductor and, when the field is
applied this charge can flow freely across the semiconductor. Thus,
there is no rectifying behaviour in this case. If, on the other hand, the

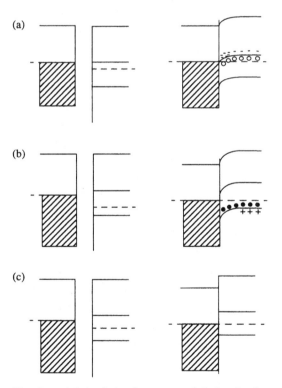

Fig. 5.1 *Schematic band structure of the barrier formation between a low work
function metal and (a) an n-type semiconductor, (b) a p-type semiconductor, and
(c) a pristine polymer. The open and closed circles indicate acceptors and donors, and
the crosses and bars indicate holes and electrons.*

semiconductor is p-type, the positive charge on the metal will repel the holes in the semiconductor and give rise to a depletion layer as depicted in Fig. 5.1b. When the field is applied, the dopant charges in the depletion layer will act as a barrier for conduction and the contact is rectifying.

When the semiconductor is intrinsic, i.e., absolutely undoped and with the Fermi energy exactly at midgap, no depletion layer can be formed and all electrons have to go into the conduction band of the semiconductor. As discussed below, the polymer samples that are prepared under UHV conditions are extremely pure and are very close to pristine (Fig. 5.1c) and thus also very poor conductors. This is also because of the fact that the undoped polymer is a molecular solid and that inter-molecular interactions are rather weak. The charge is therefore transferred from the metal to the first monolayer of the polymer only[7]. If the conduction band edge lies considerably above the Fermi energy of the metal, we can also have the situation of no charge transfer, i.e., the semiconductor behaves as an insulator. Since no charge is injected into the bulk polymer in either of these cases, the electronic bands remain planar, i.e., there is no band bending in contrast to the situation in inorganic semiconductors, where the depletion layer results in a quadratic bending of the bands. Thus, the prerequisites for charge transport across the interface are fundamentally different in metal–polymer interfaces as compared to metal–semiconductor interfaces. This will be further discussed in chapter 8 below.

5.3   Polymer light emitting diodes

In Fig. 5.2 is shown the effect on the band bending for increasing forward bias voltage across the LED. The case depicted in Fig. 5.2a corresponds to zero bias. The voltage drop is assumed to take place over the whole polymer layer in accordance with the absence of charge in the material. The metal on the righthand side of the polymer is a high work function metal which serves as a hole-injecting contact. Note, however, that there is in principle no restriction in terms of the work functions of the two contact metals in order for a current to pass through the device. For instance, if the electron-injecting contact is replaced by a high work function metal, no electrons will be injected into the polymer and the device becomes hole-conducting only[8]. The current/voltage *(I/V)* characteristics are determined by majority carriers, in this case holes. In order for light emission to occur, however, there must be a balance between the electron and hole injection, since it is only when e-h pairs recombine that light can be emitted. Therefore, the luminescence efficiency is determined by the number of minority carriers. The choice of work function for the metals sketched in Fig. 5.2 is such that electron and hole are injected with approximately the same probability.

When the device is biased forward, the voltage drop across the polymer is compensated. In the case when the applied voltage equals the difference in the work functions of the two metals, the so called flat band condition is obtained (see Fig. 5.2b). When the applied voltage exceeds this value, the width of the potential barrier for charge injection decreases and, at some critical field, charge injection into the polymer becomes possible.

In an ideal Schottky barrier, the width of the barrier is controlled by the width of the depletion layer (i.e., the concentration of dopants). The limiting factor in this case is the barrier height.

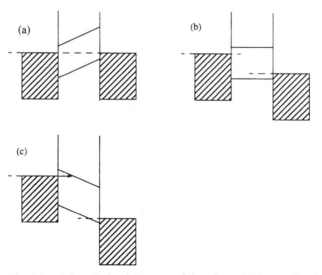

Fig. 5.2 *Schematic band structure of the polymer LED as a function of the applied voltage (a) zero bias, (b) flat band condition, (c) conducting state.*

Assuming a Maxwellian distribution of the electron velocities, some electrons, so-called hot electrons, are energetic enough to overcome the barrier and contribute to the current. The current increases in this case exponentially with temperature. This is referred to as the thermionic process and is the dominant charge-injection mechanism for the ideal Schottky barrier. Studies of, for example, PPV/ITO and PPV/Ca interfaces have shown that there is only a very weak temperature dependence in the current, in disagreement with the thermionic emission model[8]. Instead, the I/V characteristic is controlled by the electric field. The electric field sets the width of the barrier and allows for tunnelling through this barrier above a certain critical field (see Fig. 5.2). The current characteristics fit rather well to tunnelling theory, which follows the Fowler–Nordheim model[9]:

$$I \propto F^2 \exp\left(\frac{-\kappa}{F}\right) \tag{5.1}$$

where I is the current, F is the electric field strength, and κ is a parameter that depends on the barrier height, φ, and the effective mass of the charge carriers[8,10] according to,

$$\kappa = \frac{8\pi\sqrt{2m\varphi^3}}{3qh} \tag{5.2}$$

It is clear from this relation that the work functions of the device materials, which determine the barrier height, play an important role in the device characteristics. For instance, for a given polymer the efficiency for electron injection depends on the band off-set to the metal work function. This off-set is identical to the barrier height in this simplified model. A detailed study of this relation was given by Parker[8] showing a device efficiency (photons/ injected hole) varying from 10^{-6} for Au, which has a band off-set to the conduction band of MEH-PPV of about 2.4 eV, to 3×10^{-3} for Ca, which has an off-set of 0.1 eV to the polymer conduction band. Due to the exponential decrease in the current with increasing barrier (band off-set) the efficiency of the device depends critically on the choice of contact metals.

It should be noted that the rigid band model and the tunnelling process discussed above are an idealization of the real device. It is unlikely that the barrier is exactly triangular as sketched in Fig. 5.2. The results presented in this book are aimed to give a further insight into the microscopic features of the metal–polymer interfaces and how these can be related to the macroscopic models such as the relations above.

5.4 References

1. W. Schottky, R. Strömer and F. Waibel, *Hochfrequenztech.* 37, 162 (1931).
2. E. H. Roderick and R. H. Williams, *Metal-Semiconductor Contacts,* 2nd Edition (Clarendon Press, Oxford, 1988).
3. F. Garnier, G. Horowitz, X. Peng and D. Fichou, *Adv. Mater.* 2, 592 (1990).
4. J. H. Burroughes, D.D.C. Bradley, A. R. Brown, R. N. Marks, K. Mackay, R. H. Friend, P L. Burns and A. B. Holmes, *Nature* 347, 539 (1990).
5. W. R. Salaneck, unpublished.
6. M. Schott, in *Organic Conductors,* J.-P Farges (Ed.) (Marcel Dekker, Inc., New York, 1994) Chapter 12.
7. J. E. Rowe, P. Rudolf, L. H. Tjeng, R. A. Malic, G. Meigs, C. T. Chen, J. Chen, and E. W. Plummer, *Int. J. Mod. Phys.* B 6, 3909 (1992).
8. I. Parker, *J. Appl. Phys.* 75, 1656 (1994).
9. R. H. Fowler and L. Nordheim, *Proc. R. Soc. Lond. A* 119, 173 (1928).
10. S. Sze, *Physics of Semiconductor Devices* (Wiley, New York, 1981).

- *Chapter 6*

Optical absorption and emission in conjugated oligomers and polymers

In this section, the basic features of light absorption and emission (luminescence) processes in conjugated systems are reviewed. The discussion will focus on poly(p-phenylenevinylene), PPV, compounds, which provide typical examples of the physical phenomena to be highlighted in the context of polymer-based light emitting devices.

6.1 Optical absorption

In most cases, conjugated polymers present broad, inhomogeneous optical absorption bands that become more resolved when the material can be prepared in a more ordered way. This broadness originates in a number of features such as distribution of chain lengths, presence of chain bends, chain twists, or other defects, which contribute to the determination of an effective conjugation length. In the best-ordered PPV samples, the extent of effective conjugation length is estimated to be on the order of 15–20 phenylene vinylene units, by making comparison among a series of well-defined phenylene vinylene oligomers and PPV-based polymers.

Another most notable aspect relates to the vibronic coupling, that is the coupling between electronic excitations and vibrational modes. As has been stressed many times in the literature, much of the rich and fascinating physics of conjugated polymers is based on the strong connection between the electronic structure and the geometric structure: any electronic process (be it photoexcitation or charge transfer upon reduction, oxidation, or protonation i.e., doping, of the polymer chains) results in significant geometry relaxations that in turn modify the electronic properties[1-3]. Actually, this is the reason why the excitons that are thought to be formed upon excitation from the ground state S_0 to the first singlet excited state S_1, are usually referred to as polaron-excitons: the polaron part of the terminology indicates that there occurs a geometry relaxation around the electron–hole pair; the exciton terminology implies that the excited electron and the hole are not fully free to move independently from one another but are to some extent bound through Coulombic interactions. Note that in this context, the binding energy of the polaron-exciton corresponds to the difference between the total energy of the polaron-exciton and the total energy of two isolated polarons (positive polaron for the hole and negative polaron for the electron).

The way vibronic couplings modulate the lineshape of absorption spectra in PPV systems has been addressed in several studies[4-6] and adequately modelled with the help of various theoretical approaches based on the Franck–Condon approximation[7-8]. Experimentally, the influence of vibronic coupling has been most clearly seen in emission spectra[9] where, as will be discussed below, the inhomogeneous broadening of the peaks is reduced due to migration of the excitons to better ordered segments that lie at lower energy and act as emission sites[10]. However, a useful insight into the fine details of the electronic structure can also be provided by analysis of the absorption spectra of thin films of oligomers dispersed in an inert polymer matrix; these are the systems on which we now focus and that have been characterized[11], both theoretically and experimentally, by Cornil et al.: the use of well-defined oligomers allows one to characterize the systems under conditions where inhomogeneous broadening is greatly reduced.

Cornil et al. have studied the optical absorption spectra of PPV oligomers containing from two to five phenyl/phenylene rings and analysed the extent to which the vibronic couplings affect the lineshape of the spectra[11]. It is useful to set first the theoretical

approach: the simulation of the optical absorption spectra is performed
by plotting the frequency dispersion of the absorption coefficient,
which is proportional to the frequency of the incident light ω and the
imaginary part of the average polarizablity $\bar{\alpha} = 1/3 \ (\alpha_{xx} + \alpha_{yy} + \alpha_{zz})$;
the tensor components of the polarizability are calculated from a
Sum-Over-States (SOS) expression[12] where lattice relaxations are
incorporated as shown below:

$$\alpha_{ij}(-\omega;\omega) = \frac{1}{\hbar}P(i,j;-\omega,\omega)\sum_{m}\sum_{v^x}\frac{\langle g|\mu_i|m\rangle\langle m|\mu_j|g\rangle \prod_{x=1}^{3N-6} \langle 0|v^x\rangle^2}{(\omega_{mg} + v^x E_{ph}{}^x - \omega - i\Gamma)}$$

(6.1)

Here, $P(i,j;-\omega,\omega)$ is a permutation operator defined in such a way
that for any permutation of (i,j) an equivalent permutation of $(-\omega,\omega)$
is made simultaneously; ω is the frequency of incident light; i,j
correspond to the Cartesian axes x, y, and z; g denotes the ground
state and m, an excited state; v^x is the vibrational quantum number of
the x normal mode; μ_i is the i component of the dipole operator; ω_{mg}
is the 0-0 transition energy between the ground state and the m state;
$E_{ph}{}^x$ is the vibrational quantum energy of the x mode; and Γ is the
damping factor. In this case, the transition moments and transition
energies are determined with the help of the semi-empirical
Hartree–Fock Intermediate Neglect of Differential Overlap (INDO)
formalism coupled to a single configuration interaction technique
(SCI); the configurations are generated by promoting an electron
from one of the twenty highest occupied levels to one of the twenty
lowest unoccupied levels[11]. The Franck–Condon integrals, describing
the overlap between the vibrational wavefunctions, are treated
within the harmonic approximation:

$$\langle 0|v^x\rangle^2 = \frac{e^{-S_x}S_x{}^v}{v!}$$

(6.2)

where S is the Huang–Rhys factor that is related to the extent of the
modifications taking place in the excited states and roughly corresponds
to the vibrational quantum number associated with the strongest
peak in the vibronic progression. The Huang–Rhys factor is defined by:

$$S_x = \frac{M_x \omega_x}{2\hbar}(\Delta Q)^2 \tag{6.3}$$

where M_x and ω_x are the mass and angular frequency of the normal mode and ΔQ is the coordinate displacement between the ground state and the excited state. It can therefore be noted that the larger the displacement of equilibrium in the excited state, the larger the Huang–Rhys factor, and the larger the quantum vibrational number of the strongest peak in the vibronic structure.

In PPV oligomers, according to the results of site-selective fluorescence spectroscopy[13], two effective vibronic modes need to be incorporated within the SOS expression: these are taken to be at energies of 0.16 eV (1326 cm^{-1}) and 0.21 eV (1667 cm^{-1}), which correspond to the maxima of inhomogeneously broadened super-positions of a first group of individual vibronic modes at 1170 cm^{-1} (s), 1340 cm^{-1} (s) and 1500 cm^{-1} (w) and a second group at 1570 cm^{-1} (w), 1620 cm^{-1} (s), 1670 cm^{-1} (s) and 1790 cm^{-1} (s). The Huang–Rhys factor, S[total], is then distributed between these two effective modes in a 1:2 ratio in accordance with their relative intensity weightings in the fluorescence spectra. It should be emphasized that the value of S is actually not explicitly calculated but is used as a parameter in obtaining the closest fit between the simulated and measured spectra. Experimentally, low-temperature (5 K) optical absorption spectra were measured by Cornil *et al.* on solid samples consisting of dispersions of 5–10 wt. % of the oligomer in a low molecular weight polymethylmethacrylate matrix[11].

The INDO/SCI-calculated and experimentally observed optical absorption spectra of the PPV oligomers are given in Fig. 6.1; the experimental spectrum of so-called type-II standard PPV (where optimization of the synthesis and conversion procedures yields material with better order than are more generally obtained[14]) is also displayed. The comparison of the lineshapes between theory and experiment is excellent. We focus our attention on the lowest energy absorption feature, that is the S_0 to S_1 transition which primarily originates in an electron excitation between the HOMO and LUMO levels. Note that the second weak feature, appearing in the spectra of oligomers containing more than four rings, for instance around

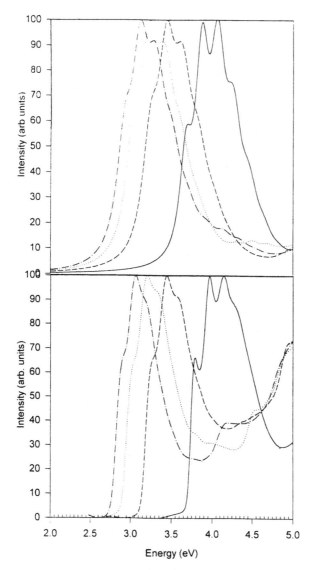

Fig. 6.1 *Experimental (lower part) and INDO/SCI-calculated (upper part) absorption spectra of the two- (solid line), three- (dashed line), four- (dotted line) and five-ring (dash-dotted line) PPV oligomers.*

4.1 eV for the five-ring compound, comes from a large mixing of configurations involving electron transitions between delocalized levels[15].

Both the experimental and calculated energies of the $1B_u$ transition are, as expected, red-shifted with increasing chain length. Furthermore, the transition energy is found to evolve linearly with the inverse number of C–C bonds along the chain; such a relation has been observed in other experimental works[16,17] and is expected from theoretical models[18,19]. This behaviour is illustrated in Fig. 6.2 where the evolution of the INDO/SCI calculated and experimental 0–1 (strongest vibronic feature) transition energies is given as a function of chain length; this plot points to the excellent agreement obtained between the two sets of values. Extrapolation of these results to the scale of large effective conjugation lengths gives an estimate of the $1B_u$ 0–1 transition on the order of 2.7 eV; this value is

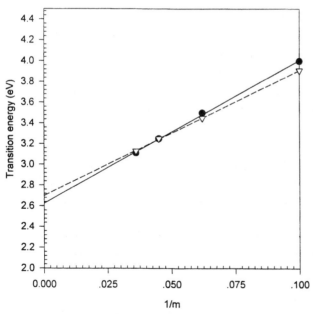

Fig. 6.2   *Evolution of the experimental (filled circles) and INDO/SCI-calculated (open triangles) 0–1 transition energies of PPV oligomers as a function of 1/m (where* m *denotes the number of carbon atoms located on the shortest pathway between the two ends of the chain).*

in very good agreement with, though slightly higher (by $\backsim 0.2\,\text{eV}$) than the experimental 0–1 transition energy of improved PPV, as seen in Fig. 6.1.

The shape of the experimental absorption spectra of the PPV oligomers, where several vibronic features are clearly observed, demonstrates that a significant change of the equilibrium position of the potential energy curve occurs when going from the ground state to the excited state. In order to match at best the experimental absorption data, the shapes of the simulated spectra are modulated by fine tuning of: (i) the size of the damping factor that controls the width of the absorption peaks; and (ii) the effective value of the Huang–Rhys factor that modifies the relative intensities of the vibronic features. The optimized parameters used when plotting the simulated spectra of the various oligomers are reported in Table 6.1.

Analysis of the experimental spectra shows that the resolution of the vibronic sidebands progressively decreases as the chain grows up; this rationalizes the different values that need to be considered for the damping factors when going from one system to the next. This behaviour appears to be intimately related to the intrinsic order in the samples; indeed, a broader distribution of conformations is expected for the longest oligomers since these systems present a large number of degrees of freedom. As a consequence, a weaker value for the damping factor is introduced when dealing with the two-ring member ($\Gamma = 0.07\,\text{eV}$); this gives rise to a simulated spectrum presenting highly resolved vibronic features, as observed experimentally[11]. The size of this parameter is increased for the other members of the series in order to broaden the vibronic

Table 6.1 *Parameters obtained by fitting the theoretical model to the experimental optical absorption spectra of the two- (PPV2), three-(PPV3), four- (PPV4), and five-(PPV5) ring oligomers of poly(p-phenylene vinylene). Γ (in eV) is the damping factor and S, the Huang–Rhys factor*

	PPV2	PPV3	PPV4	PPV5
Γ	0.09	0.12	0.12	0.12
S	1.95	1.60	1.60	1.60

sidebands and therefore follow the trends observed for the experimental linewidths.

The fit of the Huang–Rhys factor leads to almost the same value for all the oligomers ($S = 1.6$) except for stilbene where the optimized value is significantly higher ($S = 1.9$). The important point is that, on the basis of the fitting parameters reported in Table 6.1, the strength of the relaxation process taking place in the lowest-lying $B_u(S_1)$ state can be directly estimated. Indeed, the relaxation energy can be calculated as a sum of separate contributions which correspond to the energy of a normal mode weighted by the value of the Huang–Rhys factor associated to this mode. In the present case, where two effective modes at 0.14 eV and 0.2 eV are explicitly considered, the relaxation energy is given by:

$$E_{rel} = 0.16 \, S_1 + 0.21 \, S_2 \text{ with } S_2 = 2 \, S_1 \qquad (6.4)$$

Applying this sum rule to the results of the fitting procedure gives an estimate of the relaxation energy on the order of 0.34 eV and 0.29 eV for stilbene and the other oligomers, respectively. (These results are in close agreement with recent AM1 calculations[20] performed on planar PPV oligomers where the corresponding values are 0.28 eV and 0.22 eV, respectively; this confirms the reliability of the fitting parameters given in Table 6.1.) The way the Huang–Rhys factor evolves is consistent with theoretical modellings that predict this factor to decrease as the chain length grows[21]. Note that these trends are confirmed by looking at the absorption spectra of improved PPV where the highest vibronic feature is found to be no longer the 0–1 transition but rather the 0–0 transition.

At this stage, before discussing the emission aspects, we would like to stress that one has to be cautious when characterizing the relaxation phenomena in the excited states on the basis of the Stokes shift. As a matter of fact, if the same excited state is involved in the absorption and emission processes, then the Stokes shift that is defined as the difference between the 0–0 lines in absorption and emission is always negligible, as found for the PPV oligomers and in the best PPV samples. However, such behaviour is not at all inconsistent with the presence of significant relaxation processes in the excited state, as has been shown above. Note that a Stokes shift is observed either when light absorption and emission originate from two different excited states, as is the case for polyenes[22], or when random walk of the photogenerated excitons takes place[10].

6.2 Light emission

Light emission obviously constitutes the essential ingredient when dealing with polymer-based light emitting devices. The luminescence properties of conjugated oligomers and polymers have therefore been investigated in great detail. The conjugated systems can be classified into two simple categories: the compounds that luminesce and those which do not.

The absence of any significant luminescence in a conjugated oligomer or polymer is often directly related to the presence of a singlet excited state of the same symmetry as the ground state below the singlet excited state which is seen in the absorption spectra. In polyenes and polyacetylene, the $2A_g(S_1)$ state is located below the $1B_u(S_2)$ state: absorption takes place from the ground state to the S_2 state and the excitation subsequently relaxes into the S_1 state from which emission is strongly quenched[22]. This situation does not prevail in PPV compounds, for which the S_1 state has B_u symmetry and is the state responsible for the lowest energy features in both absorption and emission.

The exact relative locations of the lowest excited B_u and A_g states are difficult to predict on a theoretical basis. Indeed, they sensitively depend on the interplay between electron correlation effects and bond-length alternation effects, as shown for instance by Soos and his co-workers[23]. Strong effective bond alternations favour the $1B_u$ state as the S_1 state; this is the case in polyparaphenylene and PPV due to the presence of phenylene rings. The effective bond alternation is much weaker in polyacetylene while it is intermediate in polythiophene where the $2A_g$ state is found to lie slightly above the $1B_u$ state.

Another aspect worth mentioning is that it is usually the case that the $2A_g$ state has a covalent (neutral) character while the $1B_u$ state has a more ionic character. As a result, in polyenes end-capped by strong donor and acceptor groups, the $1B_u$ state becomes very much stabilized and passes below the $2A_g$ state[24]; substitution of the polyacetylene chains with donors and acceptors could therefore be a way to enhance the luminescence in such compounds.

Strong luminescence is thus observed in PPV and related compounds following either photoexcitation or charge injection. The photoluminescence and electroluminescence features are very similar[25], which indicates that the same species, singlet $1B_u$ polaron-excitons, are responsible in both cases for the emission processes. Note that the value of the binding energy of these polaron-excitons in PPV and related compounds remains controversial: estimates range between a few hundredths of an electronvolt to a full electronvolt with the most reasonable range being on the order of 0.1–0.4 eV[26].

While, as discussed above, improvement in sample quality and intra-chain order has a marked effect on the absorption spectra, the effect on the emission spectra is weaker. Emission spectra show better resolved structure than the absorption spectra[14], which arises from spectral diffusion within the range of accessible singlet exciton states produced by photoexcitation or charge injection. Spectral diffusion involves migration of the excited states to the lowest-energy (i.e., most conjugated) polymer chain segments prior to their radiative decay and consequent emission. Emission spectra thus sample only the most conjugated segments of the polymer chains. Absorption spectra, on the other hand, contain contributions from all polymer segments; absorption occurs instantaneously and thus samples the distribution of conjugation lengths in its totality.

Finally, it should be stressed that photoluminescence has usually a larger quantum yield in solution than in the solid state[27]. In the latter case, inter-chain interactions appear to quench the luminescence properties through a mechanism that is not fully understood yet. In this context, it is, however, useful to note that the possibility that the polaron-excitons evolve into excimers[28] is ruled out by the absence of Stokes shifts in good-quality PPV oligomer and polymer samples[14].

6.3   References

1. A. J. Heeger, S. Kivelson, J. R. Schrieffer, and W. P. Su, *Rev. Mod. Phys.* **60**, 782 (1988).

2. J. L. Brédas and G. B. Street, *Acc. Chem. Res.* **18**, 309 (1985); J. L. Brédas, *Science* **263**, 487 (1994).

3. S. Stafström, J. L. Brédas, A. J. Epstein, H. S. Woo, D. B. Tanner, W. S. Huang and A. G. MacDiarmid, *Phys. Rev. Lett.* **59**, 1464 (1987).

4. J. B. van Beek, F. Kajzar and A. C. Albrecht, *J. Chem. Phys.* **95**, 6400 (1991).

5. S. Aramaki, W. E. Torruellas, R. Zanoni and G. I. Stegeman, *Opt. Commun.* **85**, 527 (1991).

6. D. Beljonne, J. L. Brédas, M. Cha, W. E. Torruellas, G. I. Stegeman, W. H. G. Horsthuis and G. R. Mölhman, *J. Chem. Phys.*, in press.

7. Z. Shuai and J. L. Brédas, *J. Chem. Phys.* **97**, 5970 (1992).

8. Z. G. Soos and D. Mukhopadhyay, *J. Chem. Phys.* **101**, 5515 (1994).

9. D. D. C. Bradley, *Adv. Mater.* **4**, 756 (1992).

10. R. Kersting, U. Lemmer, R. F. Mahrt, K. Leo, H. Kurz, H. Bässler and E. O. Göbel, *Phys. Rev. Lett.* **70**, 3820 (1993).

11. J. Cornil, D. Beljonne, Z. Shuai, T. Hagler, I. Campbell, C. W. Spangler, K. Müllen, D. D. C. Bradley and J. L. Brédas, *Chem. Phys. Lett.*, in press.

12. B. J. Orr and J. F. Ward, *Mol. Phys.* **20**, 513 (1971).

13. S. Heun, R. F. Mahrt, A. Greiner, U. Lemmer, H. Bässler, D. A. Halliday, D. D. C. Bradley, P. L. Burn and A. B. Holmes, *J. Phys.: Condens. Matter* **5**, 247 (1993).

14. K. Pichler, D. A. Halliday, D. D. C. Bradley, P. L. Burn, R. H. Friend and A. B. Holmes, *J. Phys.: Condens. Matter* **5**, 7155 (1993).

15. J. Cornil, D. Beljonne, R. H. Friend and J. L. Brédas, *Chem. Phys. Lett.* **223**, 82 (1994).

16. M. Deussen and H. Bässler, *Synth. Met.* **54**, 49 (1993).

17. G. Horowitz, A. Yassar and H. J. von Bardeleben, *Synth. Met.* **62**, 245 (1994).

18. J. L. Brédas, R. Silbey, D. S. Boudreaux and R. R. Chance, *J. Am. Chem. Soc.* **105**, 6555 (1983).

19. P. M. Lahti, J. Obrzut and F. E. Karasz, *Macromolecules* **20**, 2023 (1987).

20. D. Beljonne, Z. Shuai, R. H. Friend and J. L. Brédas, *J. Chem. Phys.* **102**, 2042 (1995).

21. Z. Shuai, J.L. Brédas and W.P. Su, *Chem. Phys. Lett.* **228**, 301 (1994).

22. B. S. Hudson and B. E. Kohler, *Chem. Phys. Lett.* **14**, 299 (1972); B. E. Kohler, *Chem. Rev.* **93**, 41 (1993).

23. Z. G. Soos, D. S. Galvao and S. Etemad, *Adv. Mat.* **6**, 280 (1994).

24. D. Beljonne, F. Meyers and J. L. Brédas, submitted for publication.

25. A. R. Brown, D. D. C. Bradley, J. H. Burroughes, R. H. Friend, N. C. Greenham, P. L. Burn, A. B. Holmes and A. Kraft, *Appl. Phys. Lett.* **61**, 2793 (1992).

26. E. L. Frankevich, A. A. Lymarev, I. Sokolik, F. E. Karasz, S. Blumstengel, R. H. Baughman and H. H. Hörhold, *Phys. Rev. B* **46**, 9320 (1992); P. Gomes da Costa and E. M. Conwell, *Phys. Rev. B* **48**, 1993 (1993).

27. J. Grüner, P. J. Hamer, R. H. Friend, H. J. Huber, U. Scherf and A. B. Holmes, *Adv. Mat.* **6**, 748 (1994).

28. S. A. Jenekhe and J. A. Osaheni, *Science* **265**, 765 (1994).

- *Chapter 7*

Examples

7.1 Background

In this chapter, the chemical and electronic structure of the interfaces between low work function metals and prototypical conjugated oligomers and polymers, appropriate for polymer LEDs, which have been studied over the past several years, are discussed. One of the underlying themes in these discussions is aimed at illustrating the combined experimental—theoretical approach, modelling the initial stages of the interface formation corresponding to the deposition of the metal onto the polymer surface, which has proven to be a very fruitful practice, yielding more than the simple sum of the individual parts[1–8]. Following the description of the pertinent experimental sample preparation procedures, the examples are arranged in an order convenient for developing certain themes in the logic of the descriptions of the physical phenomena. More detail is supplied in the first examples, whereas less is supplied in subsequent examples, essentially pointing out differences. Also, an order has been chosen for presentation which minimizes the necessity of repetition of certain aspects.

From the experimental standpoint, the chemical and electronic structures of the surfaces of thin films of the conjugated materials are first studied in their pristine state, in order to generate surface electronic band structure parameters for input into device performance models. Then these surfaces are studied as metal atoms are slowly deposited, as the early stages for formation of the metallic overlayer develops. Two surface-sensitive techniques are used, each of which directly represents somewhat different but related aspects of the electronic structure of the polymer. X-ray photoelectron spectroscopy (XPS), which probes the binding energies of the core levels, is used to:

■ indicate the chemical state (purity, oxygen contamination, etc.) of the as-prepared surface;
■ provide a measure of the degree of metal atom coverage, step by step;
■ indicate the ionic state of the metal atoms; and
■ provide a measure of any tendencies of the metal atoms to cluster on the surface and/or diffuse into the near surface region.

On the other hand, ultraviolet photoelectron spectroscopy (UPS) is used to:

■ study the density of valence electronic states of the pristine material;
■ provide electron energy band information at the surface;
■ provide a reasonable measure of the sample work function as well as changes thereof during metal atom deposition; and
■ provide a good measure in the changes in the density-of-valence-states (DOVS) as a function of the degree of metallization of the surfaces, at least during the early stages of interface formation, during which the low energy photoelectrons from the substrate still may traverse the metallic overlayer without undue attenuation of the signal intensity associated with electron energy loss events in the overlayer.

It is important to stress that these techniques are non-destructive for almost all organic systems, and that no indications of sample damage during the accumulation of spectra have ever been observed in the course of the measurements reported here. In each case described below, the experimental results are compared to the results of theoretical quantum-chemical studies performed on model systems for the surface and the interface of interest. The systems investigated theoretically are either the polymer material itself, modeled using isolated polymer chains, or conjugated oligomer molecules; and polymer molecules or oligomers interacting with one or more (up to a few) metal atoms. Only in a very limited number of instances has it been possible to model theoretically polymer–(metal atom)–polymer, or oligomer–(metal atom)–oligomer species. The geometries of the polymer molecules, the oligomer molecules and the complexes between the metal atoms and the organic molecules are determined with appropriate quantum-chemical methods, and the structures obtained are considered as models for the chemical species present at the experimentally studied interfaces It is important to note at the outset that *vapor-deposited molecular solid films, of well chosen model molecules, can serve as spectroscopically ideal starting points for studies of the initial stages of metal-on-polymer interface formation*[9].

Unless stated otherwise under each example, the following experimental conditions and sample preparation procedures have been followed.

7.1.1 *Spectroscopy conditions*

Photoelectron spectroscopic studies were performed in an ultra-high-vacuum (UHV) photoelectron spectrometer, the details of which can be found elsewhere[2,10]. XPS was carried out using unfiltered Mg K_α radiation (1253.6 eV photons), and with an electron energy analyser resolution such that the Au($4f_{7/2}$) line would have a FWHM of 0.9 eV. UPS was carried out with a He-resonance lamp, in connection with a 2 m grazing-incidence uv-monochromator, with the usual 21.2 eV and 40. 8 eV photons. The electron energy analyser resolution was set to 0.2 eV in the XPS spectra and the HeII spectra, but to 0.1 eV in the HeI spectra.

7.1.2 *Ultra-thin polymer film preparation*

Polymer films were prepared by spin-coating, under controlled inert atmosphere, from typically a 1 mg/ml solution of the polymer in appropriate solvents. Films were generally made on aluminum (with a natural oxide of about 20 Å), or on gold. The metals were in the form of vapor-deposited films, of about 2000–3000 Å in thickness, deposited in UHV on the surfaces of optically flat Si(110) substrates. Occasionally, the bare (natural oxide) surface of the Si substrate was used directly, depending upon the preferences for film formation displayed by the particular polymer/solvent combination, or to check on results for polymer films coated on metal substrates. Typical thicknesses of the polymer layers were controlled in the range of 100 to several hundred ångströms. Freshly prepared samples were introduced into the preparation chamber of the vacuum system (base pressure, $p \leq 10^{-10}$ Torr), directly following spin-coating, but transported through atmosphere. Following stage-wise pumping down to a pressure of about 10^{-9} Torr, samples were heated briefly to about 150 °C to eliminate any residual solvent, adsorbed water vapor or possible atmospheric contamination. After returning to room temperature, the samples are moved into an analysis chamber (pressure $\sim 10^{-10}$ Torr), where XPS and UPS measurements are carried out. Characterization by XPS included: check for elemental impurities and general elemental composition; check for oxygen contamination, via O(1s) intensity and binding energies, relative to those expected; polymer integrity, via shake-up satellite intensities relative to the main C(1s) line; an estimate of the polymer film thickness, via attenuation of core electron signals from the substrate; and an indication of the presence of island formation through an (electron take-off) angle-dependence of the above spectra. Under these conditions, it was found that (within a certain yield factor) spectroscopically pure polymer surfaces could be prepared, provided that the starting material was sufficiently pure. All studies reported, unless specifically mentioned, are for well defined surfaces essentially free from contamination by oxygen-containing species, at the level of about 3 atomic % oxygen or less, as determined by XPS utilizing appropriate photoionization cross sections and analyser transmission characteristics.

7.1.3 *Ultra-thin condensed molecular solid film preparation*

First, gold substrates were prepared, as for the polymer spin-coating samples, by PVD of 2000–3000 Å of gold onto optically flat Si(110) wafers in UHV, transported in air to the PES apparatus, then cleaned by Ne^+ ion etching (sputter cleaning) before PVD of molecular solids. The gold films were found to be spectroscopically free of contamination, and exhibited the UPS spectrum, especially the sharp Fermi edge and work function, expected. Thin films of model molecules were prepared by physical vapor deposition, PVD, from a simple limited-effusion Knudsen cell, through a carefully designed system of cold (10 K) baffles, onto the sputter-cleaned gold substrates, in order to obtain as pure and controlled a molecular source as possible[2,10]. The molecular compounds, which are solids at room temperature, were contained in a borosilicate glass crucible inserted into a copper receptacle, which is gradually heated. The evaporation of the compound was monitored with a mass spectrometer, and the vapor deposition commenced (by moving the substrate in front of the molecular vapor stream) when the mass spectrum showed that the effusion from the cell consisted of only the molecules desired. The molecules were deposited at low substrate temperatures, -50 to $-150\,°C$, depending upon the vapor pressure of the molecule under study. The thicknesses of the resultant films were in the range of 50–200 Å, as determined by *in situ* monitoring of the attenuation of the core-electron signals from the substrate, $Au(4f_{7/2})$ in the case of gold. Resultant films were found by $XPS(\theta)$ to be pure, homogeneous and free of oxygen contamination. It was found that the 'oxygen contamination/impurity problem' often encountered in real polymer films could effectively be eliminated through the careful use of model molecules, as described above. A self-distillation process, which is automatically built in to the PVD process, may be partially responsible for the molecular purity and the high quality of the resultant thin films.

7.1.4 *In situ deposition of metal atoms*

Different metal atoms were deposited in different ways. Aluminum was deposited from a hot tungsten filament source, shielded by a liquid nitrogen temperature shroud, through a small hole in a cold baffle kept at 10 K. Following initial melting-in of a new aluminium source, spectroscopically pure (via XPS and UPS valence band spectra) thick (100 Å) films could be prepared to monitor the quality of the source. Subsequently, PVD parameters were calibrated in order to control the deposition rate in the mono-layer-at-a-time region, by studying, with XPS, deposition on selected substrates at low temperatures (to inhibit surface diffusion effects). Finally, these pre-determined parameters were then used to control the step-wise deposition of essentially mono-layers (average) of Al-atoms on the surfaces of samples for study *in situ*. At no time during the course of the studies reported below was any oxygen contamination observed.

Calcium atoms were deposited in much the same way as Al-atoms, with the exception that the source was a covered tantalum boat to avoid splattering Ca-clusters onto the sample. Clean Ca-films could be prepared form the source for characterisation by XPS and UPS, as a check on source purity.

Sodium or rubidium atoms were deposited in much the same way as Al- and Ca-atoms, with the exception that the atoms were obtained from SAES® getter sources (zeolites), which could be mounted, one at a time, into a heated glass shield (to maintain a stream of atoms in the correct direction in the UHV chamber), then opened in UHV and maintained clean during the course of the measurements[11]. Spectroscopically clean metal films could be prepared from these sources, as a check on source purity.

Some of the theoretical models used vary in each of the studies reported. These models were discussed in chapter 2, and the results are presented in each individual case below. In some cases, however, a detailed interpretation of the experimental UPS spectra requires the support of calculated density-of-valence-states (DOVS) curves. For that purpose, the valence effective Hamiltonian (VEH) method, which has been shown to provide accurate distributions of valence states for polymers and molecules, is used[12–14]. Even though it cannot be used for metal-containing systems (since the VEH potentials are not defined for metal atoms), the VEH approach is nevertheless of interest to determine the nature of the electronic wavefunctions (molecular orbitals) associated with a given UPS feature in the pristine polymer.

7.2 The aluminum on polythiophene interface

Although polythiophene may not be presently one of the most attractive materials for polymer-based LEDs, it has played an important role in LED development. For example, the first polymer-LED with a significant degree of polarized light was fabricated from a substituted polythiophene[15]. From the experimental interface standpoint, two of the more promising thiophene systems are poly-3-octylthiophene (P3OT) and α-sexithiophene (α6T), the geometrical structures for which are shown in Fig. 7.1. In the pristine polymer systems[3,16], the XPS C(1s) and S(2p) core level binding energies are located at −285.1 eV and −164.1 eV in P3OT, while for α6T, these peaks appear at −284.9 eV and −164.2 eV, respectively. In the polymer, the contributions of the conjugated carbon atoms and of the carbon atoms in the alkyl side chains to the C(1s) spectrum cannot be easily resolved, although indications do appear in the spectra.

The XPS C(1s) lineshape of α6T changes as aluminium atoms are deposited at room temperature: a shoulder grows on the low binding energy side of the peak (Fig. 7.2). The position of this shoulder is estimated to be at −282.6 ± 0.2 eV. It corresponds to a *new carbon species* appearing at the interface as a consequence of the interaction of aluminum atoms with the polymer. The chemical shift of this new species relative to the main line ($\Delta E \approx 2.5$ eV) indicates that the electron density on the corresponding sites is significantly larger than that of the carbon atoms in pristine α6T. It is difficult to determine directly the precise magnitude of atomic charges from XPS chemical shifts. Comparison with well-known systems, however, can provide a useful insight into the chemical situation of the new carbon species. Relative to a 'neutral' carbon atom, e.g., in alkenes or benzene, a 2.5 eV shift to higher (more negative) binding energy corresponds to the formation of a carbonyl group (C=O). Assuming that the XPS chemical shifts can be interpreted as being directly related to the atomic charge density, the chemical modification observed for a part of the α6T carbon atoms upon interaction with

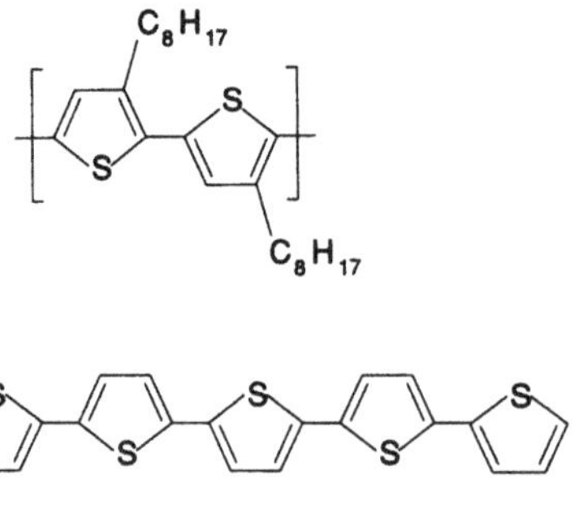

Fig. 7.1 *The geometrical structures of poly-3-octylthiophene, or P3OT (above), and α-sexithiophene, or α6T (below).*

aluminum is as strong as the transformation of a -CH=CH- group into a C=O group. The appearance of new chemical species giving rise to a C(1s) signal at low binding energies has also been observed in interfaces of aluminum with other polymers, such as polyimides and polyesters[17-19]. In those cases, aluminum atoms are thought to react primarily with heteroatoms, mostly oxygen atoms, and the new carbon groups appear to be formed only at higher metal coverages after the oxygen sites are saturated.

It is to be noticed that only a very weak shoulder is found in the C(1s) line for the aluminum on P3OT system. This is because the C(1s) spectrum is dominated by the aliphatic carbon atoms of the side chains, while aluminum is expected to preferentially interact with the carbon atoms of the conjugated backbone. When aluminum is vapor-deposited onto polyethylene[20], the C(1s) line of the polymer is not affected, reflecting the fact that only weak interactions exist between the metal and this saturated polymer. Therefore, it may be reasonably assumed that the alkyl side chains of P3OT are not strongly influenced by the aluminum deposition, and that the corresponding XPS signal remains unaffected. The aliphatic carbon atoms, however, represent the major contribution to the C(1s) line. In P3OT, the ratio of aliphatic to aromatic carbon atoms is 2:1. This ratio, combined with the fact that not all of the aromatic carbon

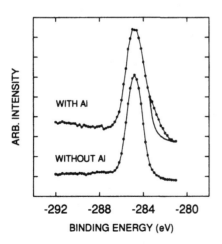

Fig. 7.2   *The XPS C(1s) spectra for step-wise deposition of aluminum on α6T.*

atoms are affected by aluminum, implies that the contribution of the new carbon species may be too small to be detected easily in the aluminum on P3OT case. The weak shoulder is observed only at high levels of aluminum coverage.

The evolution of the S(2p) line upon aluminum deposition is shown in Fig. 7.3 for P3OT (a similar evolution is observed for α6T). The apparent shoulder on the high energy side of the peak in the spectrum for the pristine material arises from the spin-orbit coupling in the S(2p) level[21]. The spectrum of the pristine system (bottom curve) indicates that only one type of sulphur site is present in each material. The appearance of the spin-splitting may be taken as an indication of the homogeneity of the sample (at least in the near surface region studied), since obviously even the slightest in-homogeneity, which would lead to slightly different apparent S(2p) binding energies, would obscure the feature. In addition, the effect of the deposition of aluminum clearly can be observed; as in the case of the C(1s) spectrum in α6T, *a new component grows* on the low binding energy side of the S(2p) line. The evolution of the spectrum as a

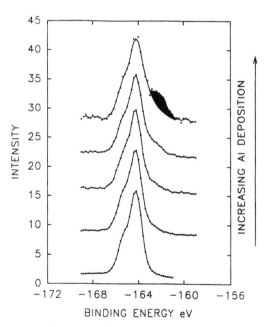

Fig. 7.3   *The XPS S(2p) spectra of aluminium on P3OT.*

function of the aluminum deposition is very similar for the two systems. The shoulder representing the aluminum-induced contribution lies at approximately -162.5 ± 0.2 eV, i.e., at about 1.6 eV lower in binding energy than the main line. Again this new feature on the high binding energy side of the main line corresponds to an increase of the overall electron density on the newly created sulphur sites, relative to the pristine thiophene system. The binding energy of the new species is close to that found in the literature for metal sulphides (Na_2S: -162.0 eV[22]; GaS: -162.2 eV[23]; ZnS: -161.7 eV[24]). By comparison, these results indicate that some of the sulphur atoms are strongly modified by the presence of aluminum, as is also the case for some of the carbon atoms.

The Al(2p) spectra for aluminum on P3OT, as a function of the coverage, are shown in Fig. 7.4. These spectra exhibit two interesting features. First, the main Al(2p) line gradually shifts towards lower binding energy as the thickness of the aluminum layer increases. This is a consequence of the improved core-hole screening for

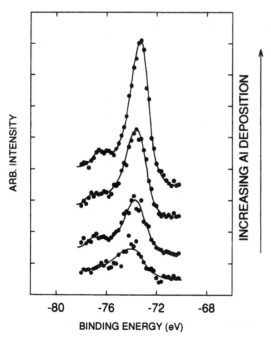

Fig. 7.4   *The XPS Al(2p) spectra of aluminium on P3OT.*

thicker aluminum deposits. In general, this shift is considered evidence that the metal is growing cluster-like on the surface, larger clusters providing stronger electronic screening than smaller clusters. The binding energy of the Al(2p) line, at -74.0 eV when first detectable following the initial deposition, shifts to -73.3 eV following successive depositions equivalent to a few ångströms (average) thickness. The value of -73.3 eV is still significantly lower (more negative) than the value for the surface of the bulk metal (-72.7 eV), indicating that the electronic properties for this layer (of clusters) are not yet those of metallic aluminum. The second important characteristic of the Al(2p) spectra is the presence of a very low intensity component at high binding energies. This component corresponds to aluminum atoms with lower electron density relative to 'metallic' atoms (corresponding to Al atoms where the electron densities are lower relative to unaffected atoms). A small amount of oxygen was detected in the P3OT films, about one oxygen per ten thiophene rings, indicating[3,25] that this component might be due to oxidized aluminum.

Combined with the evolution of the C(1s) and S(2p) spectra, the Al(2p) spectra indicate that the interaction of aluminum atoms with the conjugated backbone leads to a *significant charge transfer from the metal atoms to certain carbon and sulphur sites*. More quantitative aspects of this reaction, such as its stoichiometry, can not easily be deduced from the XPS data, because of (i) the difficulty of accurately estimating the intensity corresponding to the new species (hence their number), and (ii) the fact that with a probe depth of several tens of ångströms, the C(1s) and S(2p) signals contain contributions from areas of the organic films relatively far from the surface, where the aluminum concentration is ostensibly very small. The situation can be complicated further by the diffusion of aluminum into the organic layer. In a related study, aluminum atoms were found to diffuse into the near surface region (~ 20–30 Å) of PPV[26].

A detailed analysis of the UPS spectra of pristine P3HT, (where H = hexyl) based on the results of VEH quantum-chemical calculations of the density-of-valence-states, the DOVS, has been presented previously[27]. For the present purposes, the UPS results on P3HT and P3OT are the same; the differences in the length of the aliphatic side chains not affecting the π-electronic portion of the electronic structure. The σ-states associated with the alkyl chain predominate in the high binding energy region (for $E_B < -5$ eV, relative to the Fermi level), whereas the low energy region (from 0 eV, i.e., the Fermi level, to −5 eV) contains only contributions from the π-states on the conjugated backbone. The π-contributions, which also are present in the same energy region for α6T, are most relevant in relation with the electronic properties of these conjugated systems. The evolution of this spectral region upon aluminum deposition on P3OT is shown in Fig. 7.5. In the pristine material, the band located at 3.8 eV is related to localized electronic states with strong contributions from the sulphur atoms, while the highest occupied π-band, vanishing around −1 eV, corresponds to delocalized states with contributions from all carbon atoms, but not from sulphur atoms. The former (−3.8 eV) band appears to shift slightly toward lower binding energy as the aluminum deposition is increased, which is consistent with the fact that sulphur atoms are affected by the presence of aluminum. The behaviour of the highest occupied band

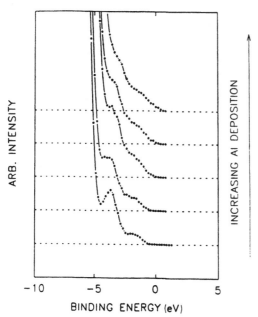

Fig. 7.5   *Evolution of the π-bands UPS spectra of P3OT with Al-deposition.*

is more difficult to assess, since the aluminum density of states probably starts to contribute to the signal near the Fermi energy for higher levels of deposition of aluminum. Note, however, the absence of a sharp Fermi edge, indicating that the electronic structure of the aluminum (if this is the dominant contribution to the UPS signal at the Fermi energy) is still not equivalent to that of the bulk metal. Nevertheless, from these spectra, it appears that the highest occupied π-levels are strongly affected by the interaction of the polymer with the incoming aluminum atoms.

The aluminum deposition also induces dramatic changes in the

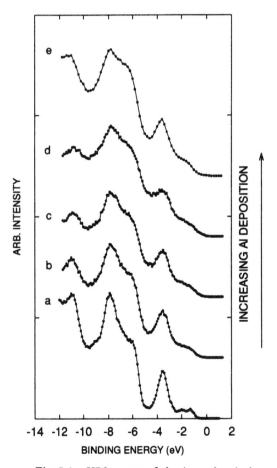

Fig. 7.6 *UPS spectra of aluminum deposited step-wise on α6T.*

UPS spectra of α6T, as shown in Fig. 7.6. The density-of-valence-states of pristine α6T (bottom spectrum) still possesses a molecular character; for instance, three features are present in the low binding energy region, at -1.3 eV, -2.0 eV and -2.9 eV (shoulder), whereas the density of states of the polymer is completely smooth in the same region. These structures disappear as soon as aluminum is present on the surface. At higher aluminum coverage, the intense band at -3.8 eV broadens and the signal around -2 to -3 eV increases, giving rise to a lineshape very similar to that observed in aluminum on P3OT (Fig. 7.5). Finally, the density of states extends towards the Fermi level.

The evolution of the inner part of the valence spectra (from -5 eV) is mostly characterized by a general broadening of the bands. As a consequence, low intensity features, such as the shoulder located around -9 eV, disappear at the highest coverages. This broadening can be related to the appearance of new electronic states corresponding to the reacted species or to the structural disorder induced in the surface layers by the incoming aluminum atoms and their reaction with thiophene units. Finally, the growing accumulation of aluminum on the surface eventually contributes significantly to the spectrum, due to the small probe depth of the low kinetic energy photo-generated electrons produced by the HeI radiation. This contribution cannot be determined accurately, however, since the density-of-valence-states of pure aluminum only shows a broad feature extending from about -6 to -10 eV[28].

From the results of *ab initio* quantum-chemical calculations on the interaction of aluminum atoms with α-terthiophene[1], it has been found that the aluminum atoms tend to form a bond to the α-carbon atoms (C_α) of the thiophene rings; the first two aluminum atoms binding preferentially to the α-carbons of the central ring, as illustrated in Fig. 7.7. The calculated aluminum–carbon bond length corresponds to the formation of covalent bonds. The same type of structure is derived for the aluminum on α6T complex, calculated using the semiempirical MNDO method[1]. The formation of the aluminum–carbon bonds leads to a local redistribution of the charge density, relative to the separate partners. As expected, the aluminum atoms become strongly positively charged. Electron density from aluminum is transferred to the α-carbon atoms and, by proximity, to the adjacent sulphur atoms, while the atomic charge of the β-carbons (C_β) remains constant. Similarly, the charges on the thiophene units without aluminum are not affected by the reaction of aluminum atoms on adjacent rings. Depending on the computational level, the estimated share of the negative charge going to the α-carbon atoms or to the sulphur atom is slightly different, but all of the calculations point to a large increase in electron density on these sites, and only on these sites, as a result of the reaction with aluminum.

These theoretical results discussed in the preceding paragraph are fully consistent with the experimental XPS and UPS data. Indeed, the new features appearing on the C(1s), S(2p), and Al(2p) XPS spectra constitute clear evidence of the charge transfer from aluminum to carbon and sulphur. Other model configurations, where the aluminum atoms are made to interact either only with the sulphur atoms or only with the β-carbon atoms, are calculated to be much less stable than the Al$-$C structure described above, and do not result in any electron density redistribution consistent with the XPS results. The inert character of the β-carbons, and of the adjacent thiophene units, upon reaction of aluminum with a given thiophene unit also is reflected in the spectra, as the major part of the C(1s) and S(2p) contribution remains unshifted when aluminum is deposited

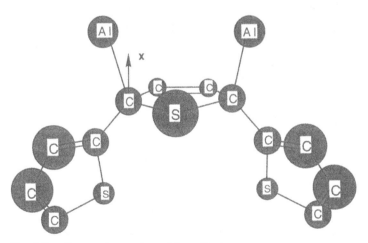

Fig. 7.7 *Al-atoms interacting with terthiophene.*

on the surface. This selectivity of aluminum bonding, i.e., the preferential reaction with the α-carbons relative to the β-carbons and the sulphur atoms, can be understood in terms of the electronic structure of the lowest unoccupied states of the pristine molecules. These states have been found to hybridize with the aluminum valence levels to form the highest occupied states of the complex. Since these states have a large contribution from the C_α, it is expected that the bonding take place preferentially on these carbon atoms. Reaction of aluminum, which is an electron donor, directly on the sulphur atoms may also be less likely, due to the inherently high electron density on those sites. It thus appears that the α-carbons can best accommodate extra negative charge, consistent with the stronger acidity of the α positions of the thiophene molecule relative to the β-carbons.

As a consequence of Al$-C_\alpha$ bond formation, the α-carbons become sp³-hybridized, the $\alpha-\beta$ carbon−carbon bonds acquire a strong single-bond character, and the $\beta-\beta'$ carbon−carbon bonds shorten dramatically. The analysis of the wavefunction of the upper occupied electronic levels shows that the π-electron conjugation along the polythiophene chain is drastically reduced by the presence of the aluminum-induced defect. The theoretical calculations indicate that the charge transfer fills the lowest unoccupied molecular orbital (LUMO) of the thiophene compounds, which shifts down near the position of the highest occupied level of the pristine system. A corresponding increase in the density of states near the ionization threshold is not observed in the UPS spectra. This is probably because the UPS spectra actually represent the density of valence states *weighted by cross-section effects*. Assuming that the cross-section of a given molecular orbital (MO) can be expressed in terms of the atomic contributions to the MO and using well-known atomic cross-sections[29], the photoelectron signal intensity of this MO can be discussed on the basis of its constitution and of the atomic cross-sections. Here, the electronic levels appearing at the lowest binding energies have strong aluminum contributions. Since the Al(3s) and Al(3p) cross-sections for HeI radiation are 20−50 times smaller than the C(2p) and S(3p) values[30], the larger density of states is compensated by the lower overall cross-section. Therefore, the calculated increase in the bare density of states at low binding energies is not expected to result in a corresponding increase in the UPS signal in that region.

Finally, it should be mentioned that the results of studies of the interaction of metal atoms other than those listed above with polythiophenes have been reported[31-33]. These other metals are mainly noble metals, however, which are not presently under active consideration for use in polymer-based LEDs.

7.3   Aluminum, sodium, calcium and rubidium on a diphenyl-polyene

A severe requirement in surface investigations dealing with reactive metals, such as aluminum, sodium, calcium, or rubidium, is the absence of contaminants. In particular, the presence of oxygen-containing impurities can affect the interpretation of the experimental results, since oxygen functionalities represent strongly competitive sites for reaction with metal atoms. While it is difficult to prepare a totally oxygen-free polymer surface, the utilization of long oligomers, or other model molecules of conjugated polymers, which can be vapour deposited (PVD) in ultra-high-vacuum to form clean thin films, constitutes an ideal alternative. A phenyl-capped 14-carbon atom polyene (i.e., seven C=C double bonds), α,ω-diphenyltet-radecaheptaene (DP7, Fig. 7.8), as a model for polyacetylene and for poly(*p*-phenylenevinylene), has been shown to be particularly useful in this respect[34].

7.3.1   α,ω-Diphenyltetradecaheptaene

Some aspects of the electronic structure of DP7 need to be mentioned in order to facilitate the discussion of the metallization. The valence electronic structure of DP7, obtained with HeII

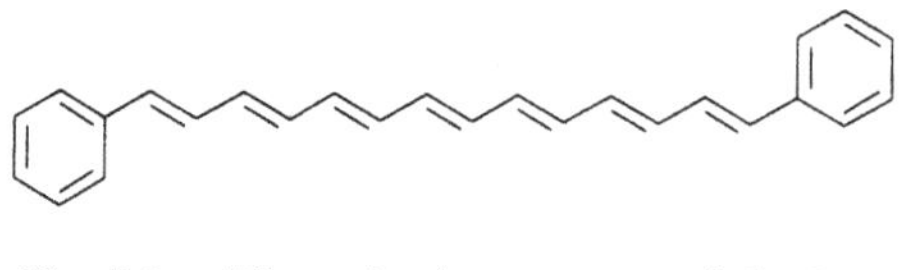

Fig. 7.8   *The molecular structure of the DP7 molecule.*

(40.8 eV) photons, is shown in Fig. 7.9, and compared with the VEH density-of-valence-states, the DOVS. Note that the overall agreement is excellent. Of particular interest is the fine-structure at low binding energies, i.e., for binding energies between about −10 and −5 eV in the UPS spectrum of Fig. 7.9. The large peak about −8 eV corresponds to electrons originating from states localized mainly (∼80%) on the phenyl rings of the molecule. The two small peaks near −3 to −4 eV, the other hand, correspond to electrons originating from states for which the wavefunctions are essentially localized (∼80%) on the polyene part of the molecule. The essentially *localized character* of these frontier molecular states, as even in shorter diphenylpolyenes[35], determines the nature of the interactions of metal atoms with the DP7 molecule.

In Fig. 7.10, is shown a portion of the C(1s) XPS spectrum for *trans*-polyacetylene, DP7, and condensed (molecular solid) benzene. The main C(1s) peaks are not displayed, but only the relatively weak satellite features that appear on their high binding energy side of the

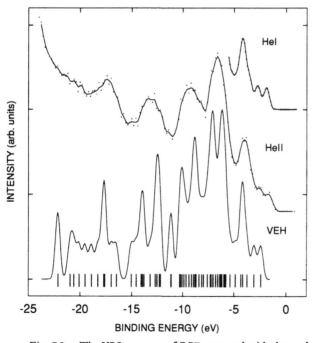

Fig. 7.9   *The UPS spectrum of DP7 compared with the results of a VEH calculation.*

main peak. These so-called shake-up (s.u.) peaks correspond to many-electron effects, which arise from valence electronic transitions simultaneous with the sudden photoionization of the C(1s) core state upon absorption of an X-ray photon, as outlined under experimental methods, above. It is clear from Fig. 7.10 that the s.u. spectrum of DP7 can be considered as the superposition of the s.u. spectra of benzene and *trans*-polyacetylene. The separation in energy of the s.u. peaks follows directly from the fact that the frontier orbitals are separated in energy into polyene-like and benzene-like. Note that relative to *trans*-polyacetylene, the major (lowest energy) s.u. peak of DP7 is at higher binding energy relative to the main C(1s) peak, and appears more intense, consistent with the fact that the optical band gap in DP7 is larger than that in polyacetylene, because the polyene chain in DP7 is shorter than the average (longer) conjugation lengths found in polyacetylene.

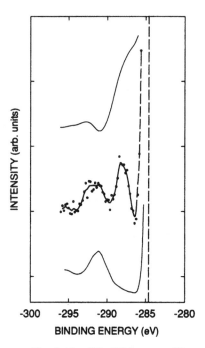

Fig. 7.10   *The XPS s.u. satellite structure on the C(1s) line of polyacetylene, DP7 and condensed benzene.*

7.3.2 *Aluminum-on-DP7 interface*

In Fig. 7.11, are presented the He (21.2 eV photons) UPS spectra obtained during the first stages of the interface formation with vapor-deposited aluminum. The strong feature in the UPS spectrum at about −8 eV (associated with molecular wavefunctions mostly localized on the phenyl end groups) is essentially unaffected by the aluminum deposition. The two weaker peaks at low binding energy (associated with molecular wavefunctions highly localized on the polyene chain), on the other hand, are strongly modified. Note again that the overall intensity decreases, as the aluminum coverage increases; this indicates the formation of an aluminum overlayer, since the cross section of aluminum is lower than that of the molecular system studied[30], as mentioned above.

The s.u. portion of the set of C(1s) spectra, corresponding to the range of aluminum coverages in the UPS spectra of Fig. 7.11, is shown in Fig. 7.12. The s.u. feature near −288 eV, which is associated with the polyene chain, is strongly affected by the presence of aluminum, while the higher energy feature near −292 eV, associated mostly with the phenyl end groups, remains essentially intact. The evolution of the Al(2p) signal during the course of the same series of measurements indicates that the initial amount of aluminum deposited is in the monolayer coverage range. In the corresponding Density Functional Theory calculations, two aluminum atoms are allowed to interact with the tetradecaheptaene molecule. The aluminum atoms

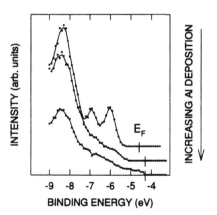

Fig. 7.11 *The UPS spectra of aluminum condensed step-wise on DP7.*

form covalent bonds with the polyene and the Al(3p) and Al(3s) orbitals hybridize with the π-molecular orbitals. The Al(3p) orbitals are found to overlap with the p_z-orbitals of two adjacent carbon atoms in the polyene, without making major changes in the planarity of the system. The double bond character in the vicinity of the aluminum atom, however, is lost.

The results presented above indicate clearly that during the initial stages of the interface formation the aluminum atoms react preferentially with the polyene chain, and not significantly with the phenyl end groups. The inert character of the phenyl ring towards aluminum deposition observed here is consistent with what is observed in other polymers. For instance, in polyethyleneterephthalate, aluminum initially reacts with the ester groups, while the phenyl groups are affected only at a later stage[18].

7.3.3   Sodium on DP7

In contrast to aluminum, where covalent reactions at the interface are observed, and a metal over-layer is then formed, it is possible to deposit sodium on the DP7 system without building a metal

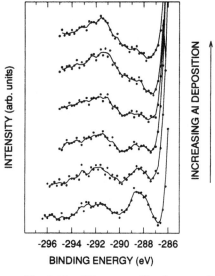

Fig. 7.12   *The s.u. satellite features of the C(1s) spectrum for aluminum deposited step-wise on DP7*

overlayer, i.e., the metal atoms diffuse easily into the bulk[36]. The XPS Na(1s) signal indicates that the Na-atoms occur in the ionic Na$^+$ state, and are uniformly distributed through the film (at least to within the extent detectable with XPS). The signal increases in intensity until the saturation value of 2:1, Na:DP7, is achieved. Thereafter, additional sodium diffuses through the sample (as studied with appropriately thin films) to the rear interface with the gold substrate, and accumulates there as sodium metal. In the UPS, changes are observed after the first few deposition steps of metal atoms.

The UPS HeI valence band for DP7 following various degrees of Na doping is shown in Fig. 7.13. The evolution of the spectrum upon doping indicates that *two new peaks*, labelled D and E, appear in the binding energy region corresponding to the original forbidden electron energy gap, E_g, in the pristine DP7, i.e., above the HOMO of

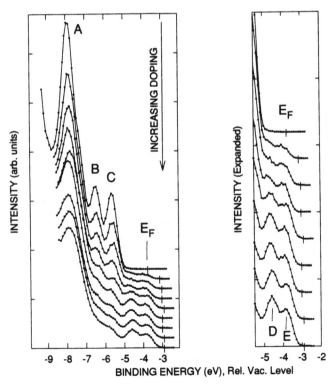

Fig. 7.13   *The UPS spectra of Na-doped DP7.*

the pristine molecule. From the theoretical modeling, these peaks have been assigned to the formation of soliton-antisoliton pairs leading to new electronic states in the otherwise forbidden energy gap between the HOMO and the LUMO of the pristine system [36,37]. The intensities of peaks B and C decrease upon increasing doping, in good agreement with the fact that the formation of a soliton-antisoliton pair leads to the removal of states from the valence band and conduction band edges of the neutral molecule to form states within the bandgap. From the XPS C(1s) and Na(1s) spectra, the doping level at which the two new peaks start to be detected is about 0.4 Na per DP7; the final (saturation) spectrum corresponds to two Na atoms per DP7 molecule[36]. The binding energy of Na(1s) is ionic, thus indicating that DP7 can accommodate two excess electrons per molecule. Geometry-optimization calculations on the $Na_2 - DP7$ complex result in the creation of two confined solitons; the bond-length alternation is completely reversed in the centre of the molecule, compared with the neutral DP7 molecule[38].

The evolution of the s.u. spectra upon sodium deposition is completely different from the case of aluminum deposition. This is shown in Fig. 7.14. In contrast to the aluminum case, the s.u. satellite related to the polyene part of the molecule does not decrease in intensity. Instead, the region between the main C(1s) peak and the polyene s.u. satellite becomes filled-in, corresponding to new low energy electronic transitions involving the new states within the previously forbidden energy gap E_g. The benzene contribution is essentially unaffected by sodium deposition, since the charge transfer resulting in the generation of the soliton pairs is mostly confined to the polyene part of the molecule.

7.3.4 *Calcium on DP7*

Calcium is used as an electron-injecting metal in PPV-based polymer-LEDs[39–41]. It was expected, therefore, that Ca would behave much as aluminum on the surface of clean PPV and model molecules there of. The valence band of DP7 as a function of increasing calcium deposition is shown[4] in Fig. 7.15. In contrast to sodium on DP7, form the results of XPS(θ) measurements, Ca seems to be localised in the near surface region of the surface. On the other hand, similar to the case of sodium, initially (at low degrees of coverage) calcium induces a decrease in the Fermi energy; two new states appear within the original HOMO–LUMO energy gap, E_g; and the atom occurs in the Ca^{2+} state. For calcium on DP7, Fig. 7.15, the doping-induced new states are closer to the Fermi level than in the case of sodium on DP7. From the doping level, as determined by XPS, the DP7 can accommodate one calcium per DP7 molecule. Since the binding energy of the Ca(2p) corresponds to Ca^{2+}, two electrons are transferred from each calcium atom to each DP7 molecule[4]. If the deposition of Ca is continued beyond the level where the DP7 surface is doped to saturation, metallic calcium appears.

The wavefunctions from the Density Functional Theory calculations on calcium on DP7 indicate that calcium-doped DP7 accommodates a bipolaron-like defect, with two excess electrons per DP7. In the modelling of an isolated molecule with two negative charges, two new occupied states appear above the HOMO of the pristine molecule[42]. When the calcium counter-ion is included, however, the electrostatic interaction of the Ca^{2+}-ion with $DP7^{2-}$ leads to the stabilization of the two new states, so that the lower state (the HOMO-1 for Ca on DP7) lies at the same energy as the HOMO of

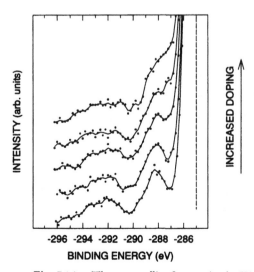

Fig. 7.14   *The s.u. satellite features in the XPS C(1s) spectra for Na-doped DP7.*

the pristine system. Therefore, only one new state (the HOMO of Ca on DP7) is expected to appear as a new feature in the original energy gap of the pristine molecule.

7.3.5   Rubidium on DP7

With the fact that sodium diffuses through DP7, and calcium diffuses into the near-surface region, it was assumed that rubidium, having a slightly larger ionic radius, might not diffuse, and thereby function as a very low work function metal electrode. Unfortunately, it was found[43] that Rb also diffuses, *albeit* somewhat slower, into the DP7 surface, and also dopes the DP7 in much the same way as sodium. On the other hand, the doping of DP7 with rubidium occurs such that at low doping levels (near 10 % or less), the formation of polarons is observed within the otherwise forbidden energy gap of DP7; while at

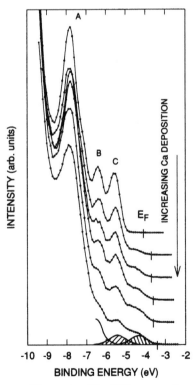

Fig. 7.15   *The UPS spectra for calcium deposited step-wise on DP7.*

higher doping, bipolarons (formally equivalent to confined soliton−antisoliton pairs in this case) are observed. Saturation occurs at the same level as for sodium, at two Rb-atoms per DP7 molecule. The UPS spectra for Rb-doping of DP7 are shown in Fig. 7.16. For rubidium, a metal with a work function less than that of sodium, the expected polaron−bipolaron transition is observed.

7.4 Aluminum, sodium, calcium and rubidium on poly(*p*-phenylenevinylene)

The detailed results of a series of studies of the interaction of metal atoms of aluminum, sodium, calcium or rubidium with the spectros-

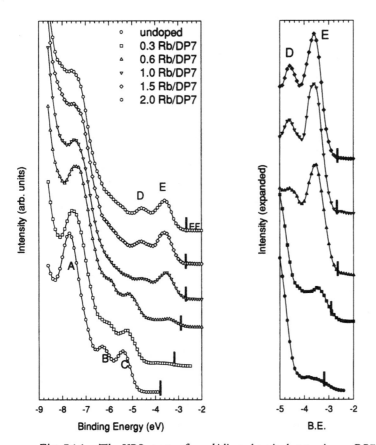

Fig. 7.16 *The UPS spectra for rubidium deposited step-wise on DP7.*

copically clean surfaces of poly(*p*-phenylenevinylene) are summarized below. In some cases, substituted PPV's were used to facilitate sample preparation of ultra-thin films by spin-coating directly the PPV polymers, rather than the precursors. Experimentally, two substituted PPVs have been used: poly(2,5-diheptyl-*p*-phenylenevinylene), or DHPPV, and poly(2-methoxy-5-(2'-ethyl-hexoxy)-*p*-phenylenevinylene), or MEHPPV, the structures for which are shown in Fig. 7.17. Surface oxygen contamination (over and above the expected oxygen content) was typically about 3 atomic % or less. In all cases, the conclusions drawn are obtained with the help of comparison with the corresponding model molecules, for which the spectra are essentially always 'nicer looking'.

7.4.1 poly(p-phenylenevinylene)

The electronic structure of poly(*p*-phenylenevinylene), or PPV, has several similarities to DP7; hence DP7 is used as a model molecule for PPV[34]. The VEH band structure of an isolated chain of PPV, and

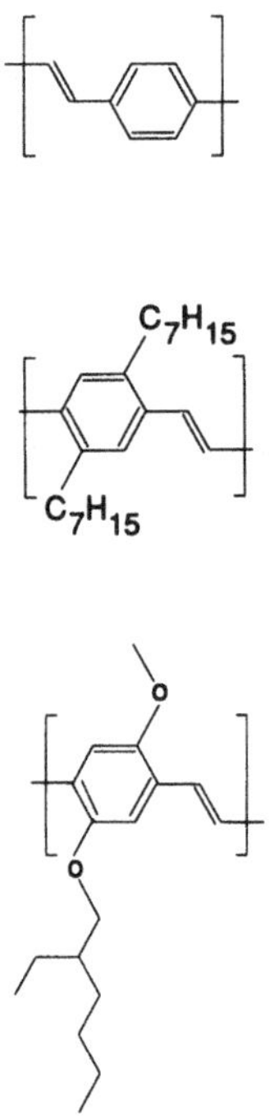

Fig. 7.17 *The molecular structures of poly(p-phenylenevinylene), or PPV, (top), poly(2,5-diheptyl-p-phenylenevinylene), or DHPPV (middle) and poly(2-methoxy-5-(2'-ethyl-hexoxy)-p-phenylenevinylene), or MEHPPV (bottom).*

the UPS spectra of the polymer, in comparison with the DOVS computed from the VEH band structure[44,45], are shown in Fig. 7.18. The PPV is taken to be planar in the VEH calculations used here, since neutron-diffraction measurements on oriented PPV at room temperature have shown that the ring torsion angles, i.e., the twist of the phenyl rings out of the vinylene plane, are on the order[46] of $7° \pm 6°$. Such small torsion angles result in negligible effects on the calculated electronic band structure compared with that for the fully coplanar conformation.

Of particular interest is the electronic structure of PPV at lowest binding energy; the region of 0 eV to about −8 eV in Fig. 7.18, where '0' in this case is the vacuum level energy. The similarity with the frontier occupied states of DP7 are as follows. In DP7, the electrons

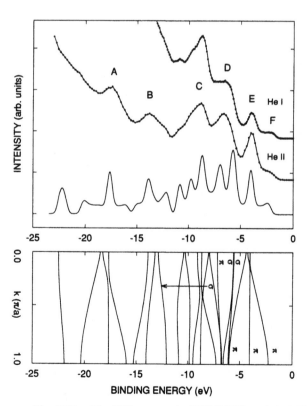

Fig. 7.18   *The UPS spectra, the VEH-DOVS, and the VEH band structure of PPV.*

in the two lowest binding energy states are in wavefunctions confined mostly to the polyene portion of the molecule, while electrons in the next three levels have wavefunctions which are localized mostly on the benzene rings. These states correspond to the first two small peaks near $-5\frac{1}{2}$ and $-6\frac{1}{2}$ eV, and the larger peak at about -8 eV, respectively, in Fig. 7.9. On the other hand, for PPV, the electrons at lowest binding energy are at the edge of the lowest energy occupied π-band, while the large peak near $-7\frac{1}{2}$ eV corresponds to electrons in a flat band, localized mostly on the benzene rings. In other words, in DP7 the lowest energy electrons are localized mostly

(a)

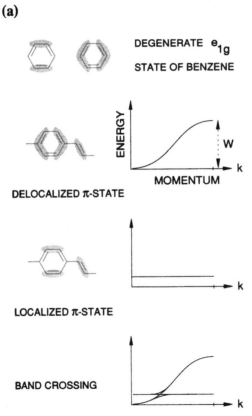

Fig. 7.19   *(a) The origin of the dispersed π-band and the flat band at lowest binding energy in the band structure of PPV from the benzene e_{1g} state; (b) the primary (substituent) and secondary (steric) effects on the combination of the dispersed and flat bands in PPV.*

to the polyene portion of the molecule; while in PPV the lowest energy electrons are in the upper portion of a π-band which is delocalized over the vinyl-groups and the benzene rings; but, *in both molecules, there are electrons localized on the benzene rings at higher binding energies* (near $-7\frac{1}{2}$ to $-8\,\mathrm{eV}$).

It is important to point out some details of the frontier band structure in pure and 2,5-substituted PPVs. A diagram of the origin of the π-band structure in terms of the lowest binding occupied e_{1g} state in the benzene molecule is shown in Fig. 7.19(a). Only one of the two degenerate halves of the wavefunction of the e_{1g} state, that which does not have a node at the point of bonding, reacts with the π-state of the vinyl group. This combination leads to the lowest binding energy dispersed π-band of PPV, shown[45] in Figs 7.18 and 7.20. It

(b)

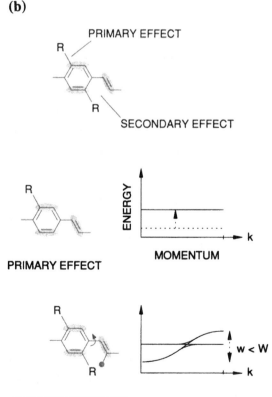

easily can be seen that differences in the dispersion of the lowest energy π-bands can be distinguished in the UPS spectra. In Fig. 7.19(b) are illustrated two effects of the 2,5-substituents: (1) the *primary* effect, where the electronic donation or withdrawal ability (charge transfer to or from the e_{1g} state) of the substituent moves the flat band up or down; and (2) the *secondary* effect, where the steric interference of the substituent and a hydrogen atom of the vinyl group leads to a twist in the backbone geometry (increase in torsion angles), resulting in an increase in the HOMO−LUMO gap. In Fig. 7.20 is shown the lowest binding energy π-band in poly(2,5-dimethoxy-*p*-phenylenevinylene), or DMeOPPV, the molecular structure of which is shown in Fig. 7.21, which illustrates both the primary and secondary substituent effects. The point to be made here is that both the primary and the secondary substituent effects can be seen and studied using UPS.

7.4.2 Aluminum on PPV

In Fig. 7.22 are shown the theoretically most energetically favourable structures of the aluminum on poly(*p*-phenylenevinylene) complex. These occur in the cases when: (a) two aluminum atoms interact with a vinylene group; and (b) two aluminum atoms interact with a

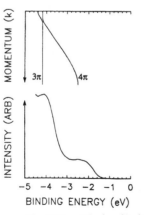 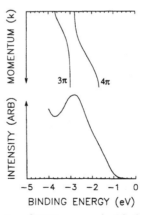

Fig. 7.20 *The low binding energy region of PPV, compared with the VEH π-band (left); and the low binding energy region of DMeOPPV, compared with the VEH π-band (right); to illustrate that these details can be distinguished in UPS spectra. The molecular structure of DMeOPPV is shown in Fig. 7.21.*

phenylene ring. The most important result is that, in full analogy to the situation found above for polyacetylene or polythiophene, each aluminum atom tends to form a covalent bond with a carbon atom[47].

The energy difference between structures (a) and (b) is, however, about 16.4 kcal per mol in favour of structure (a). The most stable configuration for the aluminum on poly(p-phenylenevinylene) complex is when the two aluminum atoms are bonded to a single vinylene group, structure (a) in the figure. This can be explained by the loss of aromaticity of the phenylene ring in case (b). The XPS and UPS studies of the initial stages of interface formation between aluminum and poly(p-phenylenevinylene) indicate, through the changes in the UPS valence band spectra and in the shake-up structure of the XPS C(1s) spectrum, that the aluminum atoms preferentially interact with the vinylene moieties. There is thus good overall agreement between theory and experiment.

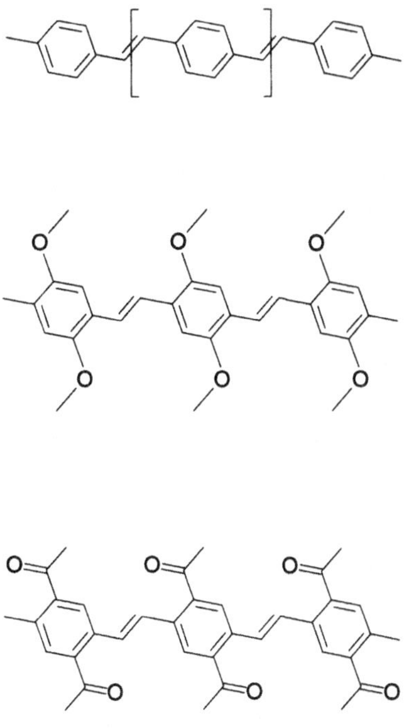

Fig. 7.21   *The molecular structures of PPV (top), poly(2,5-diformyl-p-phenylene-vinylene) (middle) and poly(2,5-dimethoxy-p-phenylenevinylene), (bottom).*

Note that, upon examination of the AM1 linear combination of atomic orbitals (LCAO) coefficients for the two aluminum on poly(*p*-phenylenevinylene) systems, the highest occupied molecular orbitals are localized in character. The HOMO and HOMO-1 levels are almost totally localized to the aluminum atoms and to the carbon atoms within the moieties to which the aluminum atoms are attached. The conjugation within the frontier orbitals is thus totally lost.

7.4.3 *Aluminum on poly(2,5-dimethoxy-p-phenylenevinylene)*

In Fig. 7.23 are displayed some of the stable structures of the aluminum on poly(2,5-dimethoxy-*p*-phenylenevinylene) complex[47]. In the case when two aluminum atoms interact with a vinylene group, structure (a), the most stable configuration is found when the two aluminum atoms are on the same side of the molecular plane. When two aluminum atoms are introduced in the vicinity of a phenylene ring, many different stable configurations are obtained, of which several are comparable in stability. Some of the most stable cases are shown, involving: in structure (b), only phenylene ring carbons; in (c), multiple-bond structure; and in (d), the formation of aluminum−O−C complex. The stabilities of structures (b) and (c) are about the same, with only 1.1 kcal per mol energy difference in favour of structure (c), while structure (d) is about 35.2 kcal per mol higher in energy than structure (c). Between structures (a) and (c),

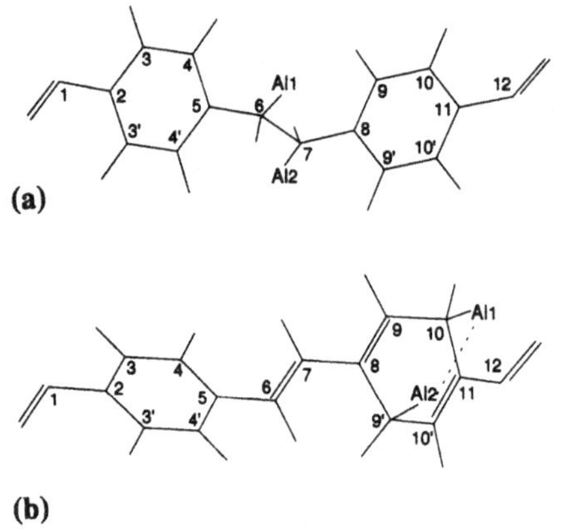

(a)

(b)

Fig. 7.22 *Interactions of Al-atoms with PPV*[74].

a)

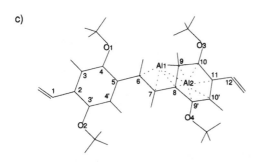

b)

c)

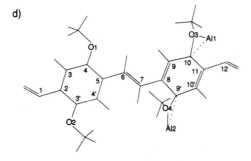

d)

Fig. 7.23   *The stable structures of the aluminum on dimethoxypoly(p-phenylene-vinylene) complex.*

there is an energy difference of about 20.2 kcal per mol in favour of structure (a). Thus, the most stable configuration is, as in the case of aluminum on poly(p-phenylenevinylene), a situation where the two aluminum atoms interact with a single vinylene group. These results indicate that the interaction with methoxy oxygens is *not* favoured in this system, and that the aluminum atoms preferentially interact with the vinylene groups in the initial stages of interface formation with poly(2,5-dimethoxy-p-phenylenevinylene).

7.4.4   Aluminum on poly(2,5-diformyl-p-phenylenevinylene)

Stable structures of the aluminum on poly(2,5-diformyl-p-phenylenevinylene) complex[47] are illustrated in Fig. 7.24. In the case where the two aluminum atoms are introduced on opposite sides of a vinylene moiety, a structure (a) is found that is similar to that in aluminum on poly(p-phenylenevinylene) with sp^3 defects on vinylene carbons. When the two aluminum atoms are found on the same side of a vinylene group, the most stable situation occurs when a carbonyl group is involved in the structure (c). The latter is more stable, but only by about 1.4 kcal per mol, than structure (a). When two aluminum atoms are introduced in the vicinity of a phenylene ring, many different stable configurations are obtained, of which several are comparable in total energy. Two of the stable structures, involving for (b) only phenylene ring carbons and for (d) the formation of an Al−O−C complex are shown in Fig. 7.24. Structure (b) is very similar to that in the corresponding situation for the aluminum on poly(2,5-dimethoxy-p-phenylenevinylene) complex; this structure is less stable than structure (c), by about 25.3 kcal per mol, due to the loss of aromaticity. Finally, structure (d) involves a C=O group as well as both a vinylene moiety and a phenylene ring; this structure is about 3.4 kcal per mol less stable than structure (c). Thus the most stable structure calculated for the aluminum on poly(2,5-diformyl-p-phenylenevinylene) complex, structure (c), involves an Al−O−C complex. This is in agreement with experimental results from studies on aluminum deposited on polymers that contain C=O groups; the results indicate that the preferential reaction sites are associated with the C=O groups[17-20].

The replacement of the methoxy groups along the dimethoxypoly(p-

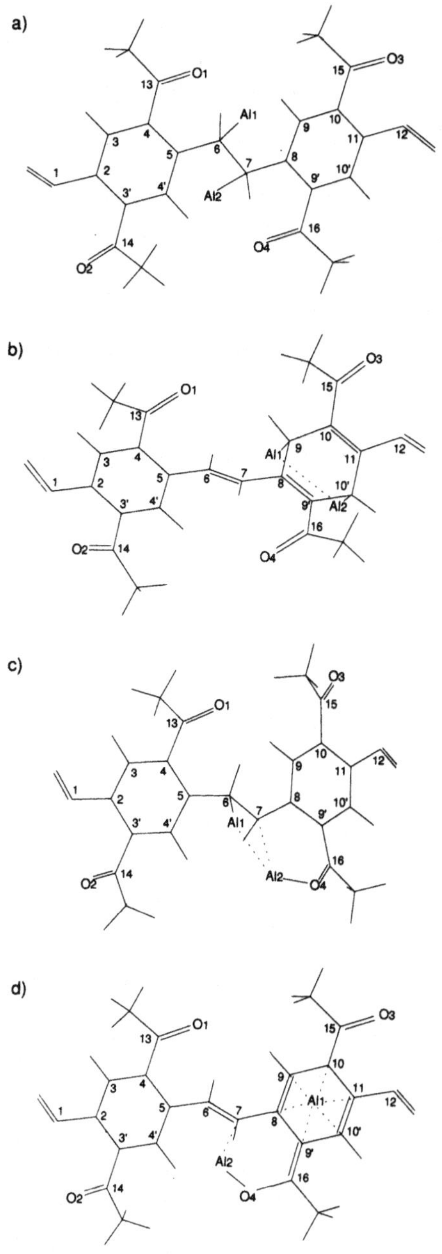

Fig. 7.24   *The stable structures of the aluminum on diformylpoly(p-phenylenevinylene) complex.*

phenylenevinylene) oligomer by the carbonyl (acceptor) groups thus allows the formation of new stable structures, which involve the appearance of Al−O−C bonds and are comparable in stability with the otherwise most stable structure, i.e., that where the two aluminum atoms interact with a single vinylene moiety. The most stable of these new structures is found when the two aluminum atoms form a complex with a vinylene moiety and an oxygen atom of a carbonyl group, as in structure (c). These results indicate that if oxygen is present in the form of C=O groups on the poly(p-phenylenevinylene) surface, the vinylene unit no longer constitutes the single most favourable site of bonding for aluminum atoms at the initial stage of interface formation. Since the oxygen atom in a carbonyl group is directly bonded to an aluminum atom in the Al−O−C complex, this is expected to be true even for other positions of the carbonyl groups.

7.4.5   Sodium on PPV

The first direct measure of multiple, resolved gap states in a doped conjugated polymer were reported by Fahlman *et al.* for sodium-doped poly(p-phenylenevinylene)[48]. Upon doping, the UPS spectra indicated a slight decrease in the workfunction at the first doping steps. At about 40% doping, defined as the Na per monomer (number of sodium atoms per polymer repeat unit) ratio, a large decrease of about 1.2 eV occurs, followed by a slight decrease as the doping level approaches 100%, i.e., one sodium atom per 'monomer' repeat unit. Simultaneous with the 1.2 eV change in the work function, i.e., at intermediate doping levels, two new states appear in the previously empty HOMO−LUMO energy gap. The intensities of the new gap states increase uniformly with the doping as shown in Fig. 7.25. The separation between the two new doping-induced peaks is about 2.0 eV at the maximum doping level, i.e., near 100%, with the lower binding energy peak positioned at about −3.2 eV. From model VEH calculations, for a 100% doping level, the new states appearing in the previously forbidden energy gap are assigned to two doping-induced bipolaron bands. Experimentally, the lack of observation of any density of states at the Fermi level, as would occur for a polaron bands, indicates the formation of bipolaron bands.

In a more recent study, the interaction between sodium and a cyano-substituted poly(dihexyloxy-*p*-phenylenevinylene), denoted by CNPPV, has been investigated[49]. The molecular structure of 'CNPPV' is shown in Fig. 7.26. Upon doping with sodium in UHV, as expected, the evolution is very reminiscent of that for the unsubstituted PPV; two new states appears in the previously forbidden energy gap, and no density of states are detected at the Fermi level, i.e., consistent with a bipolaron band formation, as can be seen in Fig. 7.27. There are some differences, however, between the results of Na-doped CNPPV and Na-doped PPV. The experimental peak-to-peak splitting of the two bipolaron peaks is about 1.05 eV in sodium-doped CNPPV, compared with about 2.0 eV for the sodium-doped PPV, as shown in Fig. 7.28. The difference in splitting between the gap states is caused by a stronger confinement of the bipolaron wavefunctions

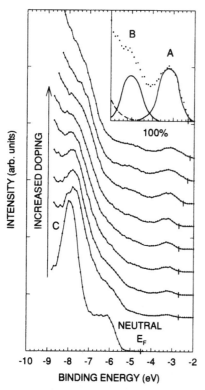

Fig. 7.25   *The UPS spectra of Na-doped PPV.*

in CNPPV. The results of AM1 calculations show that the bipolarons in CNPPV are confined on cyano–vinylene–ring–vinylene–cyano segments along the polymer backbone; the phenyl rings included in those segments can accept nearly twice as much charge as the phenyl rings outside the sequence, which are almost unperturbed[49]. The

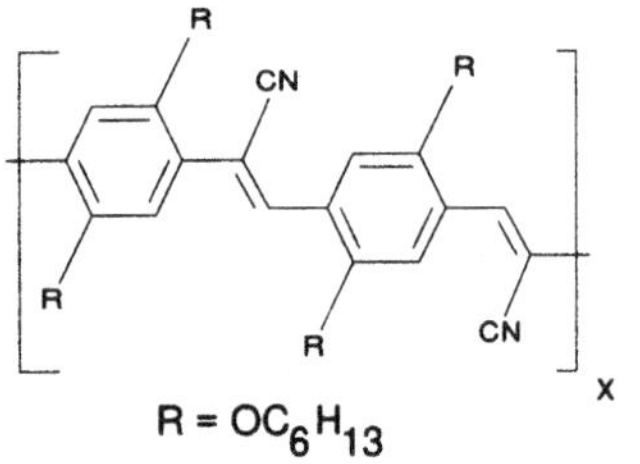

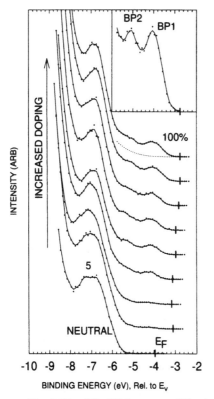

Fig. 7.26   *The molecular structure of CNPPV.*

Fig. 7.27   *The UPS spectra of Na-doped CNPPV.*

bipolaron levels occur deeper in the gap as a result of the confinement of the charge carriers.

7.4.6 *Calcium on poly(2,5-diheptyl-p-phenylene vinylene)*

The early stages of interface formation between calcium and clean surfaces of poly(2,5-diheptyl-*p*-phenylenevinylene) have been studied[50]. The results are interpreted by comparison with previously published results on a study of the doping of PPV using sodium, as discussed above. It was found that calcium does not form an abrupt interface. Instead, calcium atoms diffuse into the near-surface region of the polymer, donating electrons to the polymer, thereby becoming Ca^{2+}. The electrons transferred reside in new charge-induced self-localized states within the polymer bandgap, which appear to be bipolarons. In other words, calcium dopes the near-surface region of clean poly(2,5-diheptyl-*p*-phenylenevinylene). Clear signatures of charge transfer (doping) are distinguished in UPS spectra. Two new electronic states are induced above the π-band edge, within the otherwise forbidden HOMO–LUMO energy gap, of the pristine polymer. The energies of these new charge-transfer-induced states are near -4.0 and $-5.8\,\mathrm{eV}$ relative to the vacuum level, as shown in

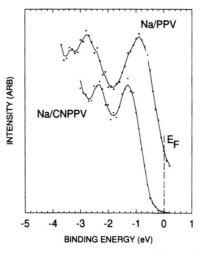

Fig. 7.28 *A comparison of the Na-doping-induced structure in the UPS spectra of PPV and CNPPV.*

Fig. 7.29. The details, the positions within the gap and the relative splitting of the two new states, are similar to those in the case of sodium on PPV[48].

In contrast to the studies of Ca deposited on essentially oxygen-free surfaces, it has been shown[51] that on surfaces containing oxygen-bearing species, oxides of calcium are formed prior to the formation of a metallic calcium electrode. The presence of an oxide interfacial layer will influence the characteristics of charge injection at such electrodes. These differences in calcium deposition, and the significance for interface characteristics will be discussed further in chapter 8. Finally, during the writing of this work, it has been observed[52] that optimum performance may be obtained on Ca-on-CNPPV-on-ITO polymer-LEDs if a certain, relatively high, background pressure of O_2 is maintained in the vapour deposition apparatus during the deposition of the calcium electrodes. Device lifetimes are improved by an order of magnitude by using 'dirty calcium' electrodes. This may serve as an example of how the basic science studies described in

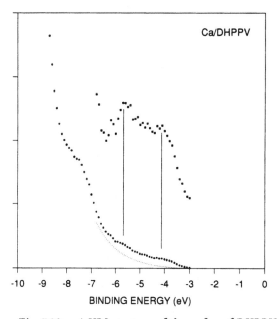

Fig. 7.29   *A UPS spectrum of the surface of DHPPV doped with calcium, showing the doping-induced bipolaron peaks.*

this text may lead to processing optimization studies which can improve device performance.

7.4.7 Rubidium on PPV

Since the rubidium atom and the rubidium ion are larger than the atom and ion of sodium, and the workfunction is less than that of sodium, rubidium was studied as a possible low workfunction electrode metal on several π-conjugated materials: poly(*para*-phenylenevinylene) (or PPV) and poly(2-methoxy-5-(2'-ethyl-hexoxy)-*p*-phenylenevinylene), (or MEHPPV), and the *trans,trans*-1,4-distyrylbenzene (or 3PV) oligomers[43]. As usual, the spectra were interpreted with the help of quantum-chemical calculations and by comparison with those of previously studied sodium-doped materials. In all of the systems analysed, a charge-transfer reaction between the metal and the polymer or oligomer occurred, resulting in n-doping

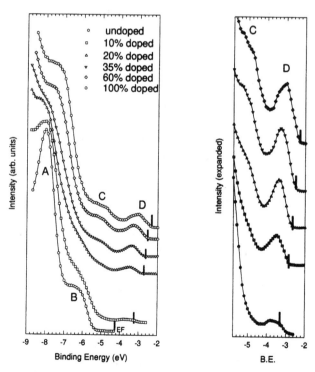

Fig. 7.30 *The UPS spectra of Rb-doped PPV.*

of the conjugated materials, as deduced by the observation of new features in the HOMO−LUMO gap and the XPS binding energy of the rubidium atoms, which was found to be ionic. As can be seen in Fig. 7.30, at high doping levels the results are similar to those obtained for the sodium-doped materials, and can be accounted for in terms of the formation of bipolaron bands in the PPVs. At low doping, however, a finite density of states is detected at the Fermi energy for all materials studied, and the spectra are interpreted in terms of polaron formation. This is the first time that a polaron-to-bipolaron transition has been observed[43] in π-conjugated polymers by UPS.

7.5 Angle-resolved UPS studies of polymers and interfaces

Although few polymer systems exist in the form of single crystals, it can be expected that in the future, polymer systems will be prepared which are more ordered. The studies reviewed here are prototypical of what can be done on ordered polymer systems, and lay the basis for future work on ordered systems using photoelectron spectroscopies.

7.5.1 4-BCMU polydiacetylene

Single crystals of polydiacetylenes (PDAs) have been studied[53,54] by X-ray photoelectron spectroscopy (XPS), without angular resolution, however, and with the low energy resolution inherent in XPS. In addition, model molecules as well as Langmuir−Blodgett and vapor-deposited films of PDAs have been studied by ultraviolet photoelectron spectroscopy, or UPS, and near edge X-ray absorption spectroscopy[55-60]. The results summarized below represent the first observation of the dispersion of the lowest binding energy π-electronic bands in a polymer, namely in 4-BCMU-PDA, using angle-resolved ultraviolet photoelectron spectroscopy (ARUPS). The molecular structure of 4-BCMU polydiacetylene is illustrated in Fig. 7.31. The polymer chains lie along the *b*-axis of the crystal, as illustrated. This particular PDA was chosen because polymer chains in poly-4-BCMU crystals are known to be very long (several micrometers), unstressed, and the inter-chain spacing is long enough that, with the short escape depths of electrons in UPS, only the top mono-layer would be studied[61,62]. Because of the large inter-chain distance parallel to the surface, the π-systems are essentially non-interacting; isolated π-chains could be studied as essentially isolated polymer chains.

The measurements were carried out using polarized-light from synchrotron radiation. The angle-resolved UPS spectra were recorded for specific directions of photon incidence, photon polarization, and electron exit, chosen in order to resolve the momentum dependence of the π-electron energy bands which could be observed in this experiment. Details are available elsewhere[63]. The UPS results are analysed not only with the help of the valence effective Hamiltonian (VEH) method, but also with the help of new quantum-chemical calculations based upon the excitation model method[64]. The full VEH band structure is shown in Fig. 7.32.

A wide (full valence band) UPS spectrum of PDA, for the electric field vector (E) parallel to the b-axis of PDA, which is the crystal axis parallel to the direction of the conjugated chains, is shown in Fig. 7.33. Because of the large number of electrons present in the large aliphatic BCMU side groups, the spectrum is dominated by the signal from the σ-system of the side groups from about $-11\,\text{eV}$ in binding energy and down. The lowest binding energy π-electron bands, however, occur at lower binding energy than the σ-manifold (of both the PDA backbone and the BCMU side chains), as noted in the figure, and can be resolved. Note, in the figure, the large intensity difference between the region of the σ-bands and the energy-resolved π-bands.

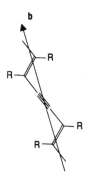

R=(CH$_2$)$_4$OCONH$_2$COOC$_4$H$_9$

Fig. 7.31 *The molecular structure of 4-BCMU PDA.*

The orientation of the crystal(s) relative to the polarization direction of the incident photon beam is a central part of ARUPS[56,65,66]. In the present work, alignment is simplified by the fact that the samples are highly one dimensional (1-D), and that the chains are parallel to the surface of the crystals. Although the orientation of the 4-BCMU PDA crystals was known, in the actual experiments the samples were aligned relative to the polarization of the photon beam by observing the dependence of the intensity of the π-band structure in the UPS spectra as indicated in Fig. 7.34. For $E//b$, the π-band edge intensity is a maximum. For $E \perp b$, the π-band intensity disappears. Thus the orientation of the samples was determined experimentally, in the UPS measurements themselves. As part of this study, it was shown how the E-vector of the light picks out the π-system, and that the E-vector of the light only excites electrons in π-states with electron momentum k for $k//E$ (or k parallel to a component of E), as shown in Fig. 7.34. Thus, by definition, *the sample itself picks out the component of the light polarization which is parallel to the direction of dispersion of the π-system* (the sample b-axis), and only a

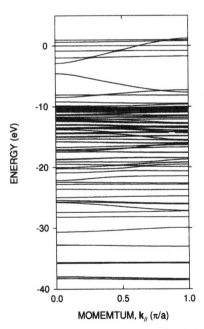

Fig. 7.32   *The VEH band structure of 4-BCMU PDA.*

decrease in intensity of the UPS signal from π-electrons is observed as the component of E parallel to the b-axis is reduced. This observation confirms that the light couples to the π-electron momentum vector, k, which is parallel to the crystal b-axis[63].

For a fixed geometry[63], the spectra are collected as a function of angle (θ) in the $k-b$ plane. The component of the momentum of the photo-emitted electron parallel to the surface, $k_{//}$, of the sample varies from zero at $\theta = 0$, and increases[66] as $k_{//} = \sin(\theta)\text{x}$ $(E_{kin})^{\frac{1}{2}} \times 0.51\ \text{Å}^{-1}$. This $k_{//}$ is just the momentum of the π-electrons of interest, and E_{kin} is the measured kinetic energy of the electrons detected. A plot of the energies of the π-electron emission peaks versus the parallel momentum $k_{//}$, in units of the width in momentum of the first BZ (π/a), yields the energy versus momentum relationship for the π-electrons. Here, a is the dimension of the unit cell of 4-BCMU-PDA in the b-direction[61,62], $4.88 \pm 0.1\ \text{Å}$. The experimental data are plotted in Figs 7.35 and 7.36, along with the π-bands derived from the VEH (Fig. 7.35) or the excitation model (Fig. 7.36) calculations. The same experimental data points are shown in each

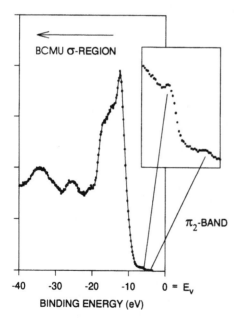

Fig. 7.33   *The full UPS spectrum of 4-BCMU PDA.*

figure. In the figures, it can be seen that the binding energy of the π_2 UPS peak changes as a function of the collection angle of the emitted electrons, θ, in units of $|k_{//}|/(\pi/a)$, according to the equation given above. Furthermore, the next higher binding energy peak does not appear to change. Because of the high kinetic energy of the photoelectrons used in these measurements, the BZ is covered by a relatively small spread in angles, hence the small number of distinguishable data points in the first BZ. Uncertainty in determination of the small angles leads to the error bars in the momentum direction, as estimated in the figures. Spectra also can be taken out at angles corresponding to the second BZ, and the results transferred (folded back) into the first zone. When this is done, the results agree with those shown in the figures.

The VEH results (Fig. 7.35) indicate that, below the dispersed π_2-band there is a band of less dispersion, corresponding to a σ-band involving the triple bonds in the back bone, and a flat (non-dispersed) band corresponding to the C=O groups in the 4-BCMU side chains. In the excitation model results (Fig. 7.36), below the π_2-band there

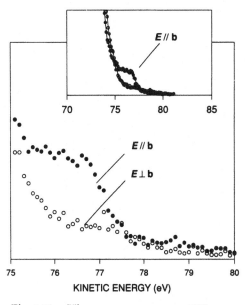

Fig. 7.34 *The narrow energy range UPS spectra of 4-BCMU PDA, for the light polarization vector, E, parallel or perpendicular to the direction of the π-system.*

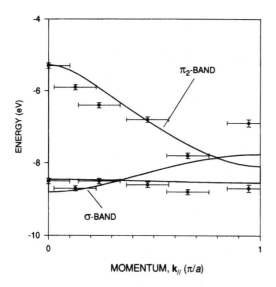

Fig. 7.35   *The ARUPS peak energies plotted versus the VEH π-band structure.*

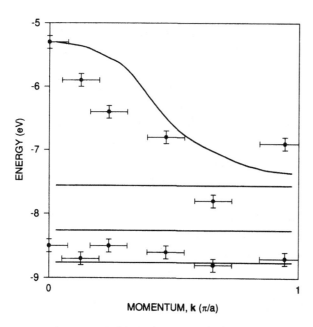

Fig. 7.36   *The ARUPS peak energies plotted versus the excitation model π-band structure.*

occur three flat bands corresponding, in order of increasing binding energy, to uncoupled triple bonds in the back bone, and to two localized C=O bands reflecting two different C=O environments in the 4-BCMU side chains. In comparing the experimental data with the results of the model calculations, it can be seen that the observed dispersion is more closely reproduced by the VEH π_2-band (Fig. 7.35). In addition, although the σ-band and the localized C=O band are not distinguishable in the spectra, all data points are consistent with the VEH theoretical model except for one point at a value of $(k_{//}a/\pi) \approx 0.95$, where the binding energy appears to be too low by almost 1 eV. In Fig. 7.36, however, using the excitation model, the calculated dispersion of the π_2-band (2.1 eV) is somewhat smaller than that in the VEH model (2.7 eV); in addition, there is no evidence in the spectra, even given the rising back ground broadness of the data near -8 eV, for the presence of three flat bands, especially the band near $-7\frac{1}{2}$ eV. This discrepancy comes mostly from the fact that, in the excitation model, the so-called perpendicular π-electron states associated with the triple bonds are considered to be completely decoupled. Because of band overlapping and the presence of at least one C=O flat band, and one σ-band with small dispersion, in the low binding energy region, the width of the π_2-band cannot be compared with the theoretical models directly.

It should be mentioned that, since data obtained at higher angles, folded back into the first BZ, do indeed fit to the band structure shown (within the error bars indicated), it does not appear that either the surface quality of the crystals or the inaccuracy in determination of the angles can account for any disagreement between the data and the calculations. The significance of the data point at $(k_{//}a/\pi) \approx 0.95$ is not clear at this time.

Although these data are preliminary because of the limitations imposed by the spectrometer used for this study[63], it is clear that the electronic structure of ordered polymer systems, such as the 4-BCMU-polydiacetylene, can be studied by angle-resolved ultraviolet photoelectron spectroscopy, and that useful information may be obtained. The coupling of the polarized light to the unique one-dimensional nature of the PDA single crystals is an essential ingredient in the study. This coupling effect, in which the *sample itself* *selects the component of polarization of the light beam parallel to the dispersion axis of the π-electrons*, enables the use of less-than-ideal geometrical configurations[63]; an effect which is utilized in studying ordering effects at polymer–polymer interfaces, below.

7.5.2   *Ordered PTFE*

Thin films of ordered parallel ribbons of polytetrafluoroethylene (Teflon®), or PTFE, have been prepared by hot-dragging techniques[67], and have been found to be highly ordered over macroscopic dimensions[68]. The electronic structure of these special films has been studied by Near Edge X-ray Absorption Spectroscopy (NEXAFS), among other reasons, as a means of determining the order of the PTFE polymer chains relative to the ribbons of material generated in the hot-dragging sample preparation procedure[69]. Although PTFE films have been prepared that are thin and uniform enough for UPS to be carried out[70], PTFE does not have π-bands at low binding energy (which are separated in energy space from the dominant σ-manifold), and thus dispersion effects were not studied.

7.5.3   *PPP on ordered polytetrafluoroethylene*

It has been shown that it is possible to obtain alignment of organic polymers and molecules by depositing the materials onto aligned polytetrafluoroethylene (PTFE) substrates[67]. This has been demonstrated in the case of poly(2,5,-p-phenylene), or DHPPP, where aligned films have been prepared by solution casting onto aligned PTFE substrates[71].

In Fig. 7.37 are shown three valence band spectra of DHPPP compared with the DOVS derived from VEH calculation for DHPPP. The inset shows the low binding energy part in more detail. The spectra have been recorded with 40 eV photon energy for three different orientations of the sample relative to the polarization direction of the incoming radiation; the incoming light (A) parallel; (B) 45°; and (C) perpendicular to the aligned chains of the PTFE substrate. From the results of the VEH calculations (D), the peak labeled 1 can be assigned to electrons removed from the two highest lying π-bands: π_3 and π_2. The π_3 band is derived from molecular orbitals delocalized along the polymer backbone and has a significant dispersion. The π_2 band is a flat band derived from molecular orbitals localized on the ortho carbons in the phenyl rings. These molecular orbitals correspond to one of the doubly degenerated e_{1g} molecular orbital in benzene, with linear combination of atomic orbital (LCAO) coefficients close to zero on the carbons connecting to the neighbouring phenyl ring (see Fig. 7.20). As seen in inset of the figure, the intensity of peak 1 is a maximum for configuration (C) and completely disappears for configuration (A), which would be expected if the polymer chains are almost completely aligned. Configuration (B) shows a spectrum for a configuration in between (A) and (C). These results indicate that (i) the DHPPP chains are ordered (parallel to one another), confirming measurements using polarized microscopy[72], but (ii) the DHPPP chains are aligned *perpendicular* to the PTFE chains (in configuration (C), the polarization of the incoming light is perpendicular to the PTFE chains but couples strongest to the dispersed band along the polymer backbone). This ordering was unexpected, but follows from the tendency of the alkyl-chains (rather than the conjugated polymer backbone) to align with the PTFE chains.

A complication, however, is the fact that the heptyl-groups make

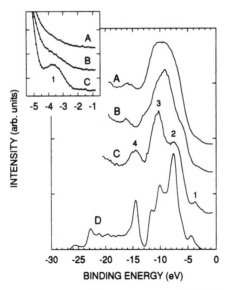

Fig. 7.37   *The ARUPS spectra of DHPPP adsorbed on the surface of ordered PTFE.*

an angle of about 30° with the direction of the conjugated polymer backbone. There also are two possible orientations of the conjugated polymers chains and their heptyl-groups, resulting in domains of each orientation on the surface of the PTFE, resulting in a certain broadness, in the sensitivity of the ARUPS signal at the π-band edge, to the angle of the E-vector relative to the orientation direction.

The above results are an example of the use of ARUPS in determining some details of polymer–polymer interfaces. To date, very few studies of polymer–polymer interfaces have been reported. As the abilities to obtain ordered polymer systems increase in the coming years, however, techniques such as ARUPS will be more and more useful in studies of such polymer–polymer interfaces[73].

7.6   References

1. M. Boman, S. Stafström and J. L. Brédas, *J. Chem. Phys.* **97**, 9144 (1992).

2. P. Dannetun, M. Lögdlund, C. Fredriksson, M. Boman, S. Stafström and W. R. Salaneck, in *Polymer-Solid Interfaces*, J. J. Pireaux, P. Bertrand and J. L. Brédas (Eds) (IOP Publishing, Bristol, 1992), p. 201.

3. P. Dannetun, M. Boman, S. Stafström, W. R. Salaneck, R. Lazzaroni, C. Fredriksson, J. L. Brédas, R. Zamboni and C. Taliani, *J. Chem. Phys.* **99**, 664 (1993).

4. P. Dannetun, M. Lögdlund, C. Fredriksson, R. Lazzaroni, C. Fauquet, S. Stafstrm, C. W. Spangler, J. L. Brédas and W. R. Salaneck, *J. Chem. Phys.* **100**, 6765 (1994).

5. C. Fredriksson and J. L. Brédas, *J. Chem. Phys.* **98**, 4253 (1993).

6. R. Lazzaroni, C. Grégoire, M. Chtaib and J. J. Pireaux, in *Polymer-Solid Interfaces*, J. J. Pireaux, P. Bertrand and J. L. Brédas (Eds) (IOP Publishing, Bristol, 1992), p. 213.

7. S. Stafström, R. Lazzaroni, M. Boman and J. L. Brédas, in *Polymer-Solid Interfaces*, J. J. Pireaux, P. Bertrand and J. L. Brédas (Eds) (IOP Publishing, Bristol, 1992), p. 189.

8. J. L. Brédas and W. R. Salaneck, in *Organic Electroluminescence*, D. D. C. Bradley and T. Tsutsui (Eds) (Cambridge University Press, Cambridge, 1995), in press.

9. W. R. Salaneck, P. Dannetun and M. Lögdlund, in *Nanostructures based on Molecular Materials,* W. Gpel and C. Ziegler (Eds) (VCH, Weinheim, 1992), p. 329.

10. M. Lögdlund, P. Dannetun, B. Sjögren, M. Boman, C. Fredriksson, S. Stafström and W. R. Salaneck, *Synth. Met.* **51**, 187 (1992).

11. M. Ramsey, *Private communication* (1993).

12. J. L. Brédas, R. R. Chance, R. Silbey, G. Nicolas and P. Durand, *J. Chem. Phys.* **75**, 255 (1981).

13. J. L. Brédas, R. R. Chance, R. Silbey, G. Nicolas and P. Durand, *J. Chem. Phys.* **77**, 371 (1982).

14. J. L. Brédas, B. Thémans, J. G. Fripiat, J. M. André and R. R. Chance, *Phys. Rev.* **B29**, 6761 (1984).

15. P. Dyreklev, M. Berggren, O. Inganäs, M. R. Andersson, O. Wennerstrm and T. Hjertberg, *Adv. Mat.* **7**, 43 (1995).

16. R. Lazzaroni, J. L. Brédas, P. Dannetun, C. Fredriksson, S. Stafström and W. R. Salaneck, *Electrochim. Acta* **39**, 235 (1994).

17. P. S. Ho, P. O. Hahn, J. W. Bartha, G. W. Rubloff, F. K. LeGoues and B. D. Silverman, *J. Vac. Sci. Tech. A* **3**, 739 (1985).

18. M. Bou, J. M. Martin, T. Le Mogne and L. Vovelle, in *Metallized Plastics,* K. L. Mittal (Eds) (Plenum, New York, 1991), p. 219.

19. C. Grégoire, P. Noel, R. Caudano and J. J. Pireaux, in *Polymer-Solid Interfaces,* J. J. Pireaux, P. Bertrand and J. L. Brédas (Eds) (IOP Publishing, Bristol, 1992), p. 225.

20. J. J. Pireaux, *Private communication* (1995).

21. K. Siegbahn, C. Nordling, G. Johansson, J. Hedman, P. F. Heden, V. Hamrin, U. Gekius, T. Bergmark, L. O. Werme, R. Manne and Y. Baer, *ESCA Applied to Free Molecules* (North-Holland, Amsterdam, 1971).

22. M. Furuyama, K. Kishi and S. Ikeda, *J. Elec. Spec.* **13**, 59 (1978).

23. J. M. Thomas, I. Adams, R. H. Williams and M. Barber, *J. Chem. Soc. Farad. Trans. II* **68**, 755 (1972).

24. C. J. Vesely and D. W. Langer, *Phys. Rev.* **B4**, 755 (1971).

25. M. Lögdlund, P. Dannetun and W. R. Salaneck, in *Handbook of Conducting Polymers,* T. Skotheim, J. Reynolds and R. Elsenbaumer (Eds) (Marcel Dekker, New York, 1996), in press.

26. P. Dannetun, Private communication. (1993).

27. W. R. Salaneck, O. Inganäs, B. Thémans, J. O. Nilsson, B. Sjögren, J.-E. Österholm, J.-L. Brédas and S. Svensson, *J. Chem. Phys.* **89**, 4613 (1988).

28. L. G. Petersson, *Solid State Commun.* **19**, 257 (1976).

29. K. Siegbahn, *J. Elec. Spec.* **5**, 3 (1974).

30. J. J. Yeh and I. Lindau, *Atomic Data and Nuclear Data Tables* **32**, 1 (1985).

31. A. Lachkar, A. Selmani, E. Sacher, M. Leclerc and R. Mokliss, *Synth. Met.* **66**, 209 (1994).

32. A. Lachkar, A. Selmani and E. Sacher, *Synth. Met.* **72**, 73 (1995).

33. A. Lachkar, A. Selmani and E. Sacher, *Synth. Met.* **72**, 81 (1995).

34. C. W. Spangler, E. G. Nickel and T. J. Hall, *Am. Chem. Soc., Div. Polym. Chem.* **28**, 219 (1987).

35. B. S. Hudson, J. N. A. Ridyard and J. Diamond, *J. Am. Chem. Soc.* **98**, 1126 (1976).

36. M. Lögdlund, P. Dannetun, S. Stafström, W. R. Salaneck, M. G. Ramsey, C. W. Spangler, C. Fredriksson and J. L. Brédas, *Phys. Rev. Lett.* **70**, 970 (1993).

37. P. Dannetun, M. Lögdlund, C. W. Spangler, J. L. Brédas and W. R. Salaneck, *J. Phys. Chem.* **98**, 2853 (1994).

38. M. Lögdlund and J. L. Brédas, *Int. J. Quant. Chem.* **28**, 481 (1994).

39. R. H. Friend, in *Conjugated Polymers and Related Materials: The Interconnection of Chemical and Electronic Structure*, W. R. Salaneck, I. Lindström and B. Rånby (Eds) (Oxford Scientific, Oxford, 1993), p. 285.

40. D. R. Baigent, N. C. Greenham, J. Grüner, R. N. Marks, R. H. Friend, S. C. Moratti and A. B. Holmes, in *Organic Materials for Electronics: Conjugated Polymer Interfaces with Metals and Semiconductors*, J. L. Brédas, W. R. Salaneck and G. Wegner (Eds) (North-Holland, Amsterdam, 1994), p. 3.

41. I. D. Parker, *J. Appl. Phys.* **75**, 1656 (1993).

42. C. Fredriksson and S. Stafström, *J. Chem. Phys.* **101**, 9137 (1994).

43. G. Iucci, K. Xing, M. Lögdlund, M. Fahlman and W. R. Salaneck, *Chem. Phys. Lett.* in press (1995).

44. M. Lögdlund, W. R. Salaneck, S. Stafström, D. D. C. Bradley, R.

H. Friend, K. E. Ziemelis, G. Froyer, D. B. Swanson, A. G. MacDiarmid, G. A. Abuckle, R. Lazzaroni, F. Meyers and J. L. Brédas, *Synth. Met.* **41–43**, 1315 (1991).

45. M. Fahlman, O. Lhost, F. Meyers, J. L. Brédas, S. C. Graham, R. H. Friend, P. L. Burn, A. B. Holmes, K. Kaeriyama, Y. Sonoda, M. Lögdlund, S. Stafström and W. R. Salaneck, *Macromolecules* **28**, 1959 (1995).

46. G. Mao, J. E. Fischer, F. E. Karasz and M. J. Winokur, *J. Chem. Phys.* **98**, 712 (1993).

47. M. Lögdlund and J. L. Brédas, *J. Chem. Phys.* **101**, 4357 (1994).

48. M. Fahlman, D. Beljonne, M. Lögdlund, P. L. Burn, A. B. Holmes, R. H. Friend, J. L. Brédas and W. R. Salaneck, *Chem. Phys. Lett.* **214**, 327 (1993).

49. M. Fahlman, P. Bröms, D. A. dos Santos, S. C. Moratti, N. Johansson, K. Xing, R. H. Friend, A. B. Holmes, J. L. Brédas and W. R. Salaneck, *J. Chem. Phys.* **102**, 8167 (1995).

50. P. Dannetun, M. Fahlman, C. Fauquet, K. Kaerijama, Y. Sonoda, R. Lazzaroni, J. L. Brédas and W. R. Salaneck, in *Organic Materials for Electronics: Conjugated Polymer Interfaces with Metals and Semiconductors,* J. L. Brédas, W. R. Salaneck and G. Wegner (Eds) (North Holland, Amsterdam, 1994), p. 113.

51. Y. Gao, K. T. Park and B. R. Hsieh, *J. Chem. Phys.* **97**, 6991 (1992).

52. P. Bröms, J. Birgersson, N. Johansson, M. Lögdlund and W. R. Salaneck, *Synth. Met.*(1995), in press.

53. D. Bloor, G. C. Stevens, P. J. Page and P. M. Williams, *Chem. Phys. Lett.* **33**, 61 (1975).

54. J. Knecht, B. Reimer and H. Bässler, *Chem. Phys. Lett.* **49**, 327 (1977).

55. K. Seki, U. Karlsson, R. Engelhardt and E. E. Koch, *Chem. Phys. Lett.* **103**, 343 (1984).

56. K. Seki, N. Ueno, U. Karlsson, R. Engelhardt and E. E. Koch, *Chem. Phys.* **105**, 247 (1986).

57. N. Ueno, S. K. N. Sato, H. Fujimoto, T. Kuramochi, K. Sugita and H. Inokuchi, *Phys. Rev.* **B41**, 1176 (1990).

58. K. Seki, I. Morisada, H. Tanaka, K. Edamatsu, M. Yoshiki, Y. Takata, T. Yokoyama and T. Ohata, *Thin Solid Films* **179**, 15 (1989).

59. K. Seki, I. Morisada, K. Edamatsu, H. Tanaka, H. Yanagi, T. Yokoyama and T. Ohta, *Physica Scripta* **41**, 172 (1990).

60. K. Seki, T. Yokoyama, T. Ohta, H. Nakahaka and K. Kududa, *Mol. Cryst. Liq. Cryst.* **218**, 609 (1992).

61. J. Berrehar, C. Lapersonne-Meyer and M. Schott, *Appl. Phys. Lett.* **48**, 630 (1986).

62. J. Berrehar, P. Hassenforder, C. Laperssone-Meyer and M. Schott, *Thin Solid Films* **190**, 181 (1990).

63. W. R. Salaneck, M. Fahlman, C. Lapersonne-Meyer, J. L. Fave, M. Schott, M. Lögdlund and J. L. Brédas, in *Organic Materials for Electronics: Conjugated Polymer Interfaces with Metals and Semiconductors,* J. L. Brédas, W. R. Salaneck and G. Wegner (Eds) (North Holland, Amsterdam, 1994), p. 309.

64. S. Pleutin and J. L. Fave, to be published

65. E. W. Plummer and W. Eberhardt, *Adv. Chem. Phys.* **49**, 533 (1982).

66. F. J. Himpsel, *Adv. Phys.* **32**, 1 (1982).

67. J. C. Whitmann and P. Smith, *Nature* **352**, 414 (1991).

68. M. Schott, in *Nobel Symposium in Chemistry: Conjugated Polymers and Related Materials; The Interconnection of Chemical and Electronic Structure,* W. R. Salaneck, I. Lundström and B. Rånby (Eds) (Oxford Scientific, Oxford, 1993), p. 377.

69. C. Ziegler, T. Schedel-Niedrig, G. Beamson, D. T. Clark, W. R. Salaneck, H. Sotobayashi and A. M. Bradshaw, *Langmuir* **10**, 4399 (1994).

70. M. Fahlman, D. T. Clark, C. Fredriksson, W. R. Salaneck, M. Lögdlund, R. Lazzaroni and J. L. Brédas, *Synth. Met.* **55-57**, 74 (1993).

71. M. Fahlman, J. Rasmusson, K. Kaeriyama, D. T. Clark, G. Beamson and W. R. Salaneck, *Synth. Met.* **66**, 123 (1994).

72. H. Wittler, *Thesis: Substituierte Poly(π-phenylene)*: Synthese, Struktur und Phasenverhalten (Mainz, 1993).

73. W. R. Salaneck and J. L. Brédas, in *Organic Materials for Electronics: Conjugated Polymer Interfaces with Metals and Semiconductors,* J. L. Brédas, W. R. Salaneck and G. Wegner (Eds) (North Holland, Amsterdam, 1994), p. 15.

74. M. Fahlman, J. L. Brédas and W. R. Salaneck, *Synth. Met.,* (1996), in press.

• *Chapter 8*

The nature of organic and molecular solid surfaces and interfaces with metals

It is perhaps obvious that the nature of the interface between a molecular solid (polymer) and a (clean) metal surface is not necessarily equivalent to the interface formed when a metal is vapor-deposited (essentially 'atom-by-atom') on to the (clean) surface of the polymer or molecular solid. Atoms of all metals are active 'in the form of individual atoms', even gold atoms. In the context of the new polymer LEDs, some of the works discussed in chapter 7 involve the study of the early stages of formation of the interface in the latter configuration (metal-on-polymer interfaces). Very little has been reported on conjugated polymer-on-metal interfaces, however, primarily because of the difficulties in preparing 'monolayers' of polymer materials on well defined metal substrates appropriate for study (via PES or any other surface sensitive spectroscopy). The issues discussed below are based upon information accumulated over two decades of involvement with the surfaces of condensed molecular solids and conjugated polymers in ultra-thin form, represented by the examples in the previous chapter.

8.1 Polymer surfaces

It is most straightforward to begin with a brief description of the very nature of the free polymer surface, following which the polymer-on-metal interface is described, before returning to additional facts about the free polymer surface. Some fundamental differences between ideal polymer surfaces and those of conventional inorganic semiconductors are outlined below[1-4].

1 The polymer surface, as in the bulk itself, may vary from partially polycrystalline to amorphous. Except for the polydiacetylenes, single crystal polymer surfaces are essentially unknown.

2 Polymer molecules relevant to polymer LEDs are large, reasonably flexible, covalently bonded chains. The polymer surfaces generally consist of gently curved sections of the polymer chains and possibly some chain ends, where all of the bonds are satisfied; in some instances, the polymer chains are preferentially parallel or perpendicular to the surface. A schematic representation of a typical polymer 'surface' is shown in Fig. 8.1. Because of surface energy effects, however, some side groups may have a tendency to be oriented preferably 'out' from the surface or 'in' towards the

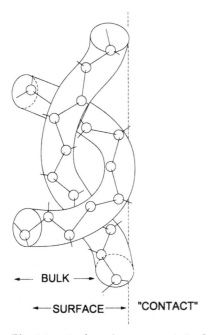

←— BULK —→

←—SURFACE—→ "CONTACT"

Fig. 8.1   *A schematic representation[3] of a polymer 'surface'.*

bulk of the film. The surface region itself is difficult to define exactly, varying from the atomic scale (~ 1 Å), to the scale of the molecular side groups (~ 10 Å), to a typical radius of gyration of a chain loop ($\sim 100-1000$ Å)[2]. The polymer chains, both in the bulk and at the surface, may assume a variety of conformations. Although electronic surface phenomena can be probed on a scale of about 10 Å, certain mechanical or chemical properties of the 'surface' may occur at larger distances in toward the bulk.

3 The truncation of the three-dimensional periodicity of a crystal lattice leads to surface state formation on essentially all inorganic semiconductor surfaces. This kind of surface state defect essentially does not occur on typical, carefully prepared, polymer surfaces. Generally, conjugated polymers consist of chemical units within which all bonds are covalently satisfied; there are no broken or dangling bonds, and the largely one-dimensional nature of the polymer chains is preserved despite (gentle) curvature and bending near the surface. On the other hand, just as inorganic semiconductor surfaces may be modified by surface treatment or preparation, the ideal nature of *intrinsic* polymer surfaces is subject to the sample preparation and treatment employed. The chemistry and treatment of polymer surfaces in general has been discussed at length[5,6]. Such effects are not considered here.

4 In principle, the interface between the metal substrate and the semiconductor or insulator can be modelled by considering the charge transferred in order to establish electrical equilibrium. In a molecular solid, the electronic states are localized essentially on the individual molecules. The charge injected into this type of material becomes much more localized than for more conventional three-dimensional semiconductors, such as silicon. The tendency for localization, via impurity scattering, electron–phonon interactions, etc., is very strong in conjugated polymers. Therefore the behavior of conjugated polymers in this case is very similar to that of molecular solids. Modeling of the electrical equilibrium at the metal–insulator (conjugated polymer) interface has been discussed recently by Schott[7]. In materials considered here, and for thicknesses of films which are appropriate for UPS studies ($\sim 20-100$ Å) or for polymer-based LED applications (typically 500–1000 Å), however,

the depletion layer distance would be large compared with the sample film thickness. Thus the flat band condition occurs[3,7].

8.2   Polymer-on-metal interfaces

Consider the polymer-on-metal interface, which might be prepared by coating a thin metal film with polymer in a polymer-based LED. The case of the counter electrode, formed by vapor-deposition, is discussed subsequently. First, assume that the substrates have 'clean' surfaces; hydrocarbon and oxide free, or 'naturally oxidized' but still hydrocarbon free (pointed out as necessary). Typically, in connection with polymer-based LEDs, the metallic substrate could be gold, ITO (indium tin oxide) coated glass, the clean natural oxide of aluminum ($\backsim 20\,\text{Å}$ in thickness), the natural oxide which forms upon freshly etched Si(110) wafers ($\backsim 10\,\text{Å}$), or possibly even a polyaniline film. 'Dirt', which may be either a problem or an advantage, will not be taken up here. Discussions will alternate between coated polymer films and condensed model molecular solid films, as necessary to illustrate points.

In the first monolayer of conjugated model material, a model molecular solid or a polymer adsorbate, assume that no chemistry (covalent bonding) occurs, since, in the absence of, for example, mechanical rupturing, the bonds at the surface of the molecular film are completely satisfied. This assumption is supported by the fact that, at least for condensed molecular solids, vapor-deposited films may be re-evaporated (removed) from the surface by gentle heating in UHV.

Without lack of generality, with the energy level (band edge) diagrams of Fig. 8.2, two cases for electrons, may be considered (there are corresponding cases for holes). If, before contact, the LUMO (Fig. 8.2) of the molecular adsorbate (E_C) lies below (at deeper binding energy) the Fermi energy ($E_C < E_F$), then electronic charge will transfer from the metal to the adsorbate; ionic 'bonding' (electronic charge transfer) will occur. If ($E_C < E_F$), then electronic charge transfer still may occur, since the image potential in connection with the metal surface can shift the energy band edges by several electronvolts[8,9]. Note that in the example, binding energies are measured positive relative to the vacuum level; therefore $E_C < E_F$ means that E_C lies deeper in energy than E_F. An example where careful and detailed measurements have been done is the case of C_{60} molecules[9–12], where the charge transfer is found to remain localized to the monolayer interface region. The condition ($E_C = E_{LUMO} < E_F$) occurs in essentially all cases of the polymer materials considered here. Since $E_{LUMO} < E_F$ because of an electrostatic effect, the charge transfer is *localized* to the first monolayer. This localization is a result of the features of the polymeric solid discussed above, namely, that the inter-chain (intermolecular) interactions are relatively weak, and, in addition to the electrostatic effects, localization occurs due to electron–phonon interactions

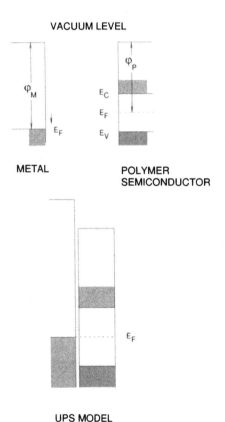

Fig. 8.2 *Metal and semiconductor surfaces shown (above); and in contact in the UPS band edge model (below).*

and/or impurity scattering. Note that in the case of pristine conjugated polymers there are no donor or acceptor states present. Therefore, no real depletion layer of electrons or holes, close to the interface, occurs, in contrast to conventional metal−semiconductor interfaces, which result in Schottky barrier formation for charge transport across the interface.

The electro-chemical potential $\mu = \mu_e + \mu_c$, consists of two terms, the electro-*static* part, μ_e, and the 'chemical' part, μ_c, as discussed in the literature[13-15]. Since the electro-chemical potential is a constant throughout the film thickness, $\Delta\mu = 0$, then $\Delta\mu_e = -\Delta\mu_c$. The electro-static part corresponds to the electro-static potential, and reflects the dipole generated by charge transfer at the interface, as seen in the step in the vacuum level in the lower part of Fig. 8.2. The 'chemical' part, in the present case, corresponds to 'the gradient in the number of particles', which is just the number of electrons transferred (per unit area) from the metal to the polymer at the interface. Since electrons are transferred to the polymer, the gradient in the number of particles works against, and just equals (cancels) the step in the electro-static potential.

For an intrinsic semiconductor with electron−hole symmetry, thermodynamic considerations indicate that the Fermi level should lie at the center of the energy gap[16,17]. It has been found experimentally that in pure molecular solid films, the Fermi energy is located at the middle of the HOMO−LUMO gap[3,9-12,18-20]. To a very good approximation, the energy difference between the Fermi energy, E_F, and the valence band edge, E_V (the ionization potential), in condensed molecular solid films is equal to one half of the optical band gap measured at the optical absorption edge; i.e., $E_g = 2(E_V - E_F)$, where both E_V and E_F are positive numbers. Thus these (pure) condensed molecular solid films appear to behave as ideal semiconductor (insulator) films[3].

A word of caution: the case of conjugated polymer films is generally identical to the case of condensed molecular solid films, as described above, but with occasional small differences. In general, the Fermi level is found to lie very near the centre of the energy gap, E_g. Small amounts of impurities, defects, or other charge donation (acceptor) species, however, can move the position of the Fermi level. For example, in the PPV materials studied in Linköping, early examples appeared to be slightly n-type. Recently, MEHPPV[21], and PPV prepared by the Cambridge precursor route[22,23], appear intrinsic[20]; $E_g = 2(E_V - E_F)$. Thus care must be taken in deducing energy band edges at the surfaces of conjugated polymers using UPS, when the Fermi level does not appear to lie at the middle of the energy gap. When the Fermi level does appear at the centre of the gap, then the UPS data agree with the optically deduced band gap.

8.3   Polymer–polymer interfaces

Interfaces between polymers or polymers and molecular solids (e.g., condensed molecular solid transport layers) also occur, for example, in polymer LEDs with electron or hole transport layers[24–26]. It appears, from within the work on polymer LEDs, that, in the absence of interfacial chemical interactions, flat band conditions apply (no band bending occurs), and that the band edge off-sets that are observed correspond to the positions of the band edges at the surfaces of the polymer components prior to contact[27]. In the absence of any mixing effects, it is expected that (conjugated) polymer–polymer interfaces in thin-film devices exhibit flat band conditions, and that the energy band edges align with each other without indications of charge-transfer-induced band edge shifts[26]. Band edge off-sets, however, are not always consistent with band edge values obtained from UPS. Differences on the order of 0.2 eV (or more) are not unusual.

8.4   The polymer LED model

In the chapter on device motivation to the study of polymer surfaces and interfaces, a diagram similar to Fig. 8.3 appears. At first glance, there is an apparent discrepancy between the *UPS model* of Fig. 8.2 (bottom), and *LED model*[22,26], in Fig. 8.3. If a second metal, with workfunction less than that of the metal shown in Fig. 8.2, is used as the counter electrode on the polymer film in the UPS model, assuming the direct alignment of the various energy band edges, the level must move, such that $\Delta\mu = 0$ throughout the sandwich, upon charge injection/transfer from the metal counter electrode to the semiconductor in the centre of the sandwich.

On the other hand, in the general procedure by which polymer LEDs are fabricated, a polymer film is coated onto a metal substrate and then a metal counter electrode is applied by physical vapor deposition. Since there is no chemistry at the polymer—metal substrate, postulated and argued above, this interface is *not* dominated by surface electronic states in the conventional semiconductor—metal contact sense. At the polymer interface with the second electrode, applied by PVD, chemistry, of one sort or another, occurs, as discussed in a wide variety of cases in chapter 7. Without loss of generality, consider the case of the device consisting of a gold substrate, MEHPPV spin-coated polymer film, covered by a calcium counter electrode applied by PVD[27]. The energy band edge relationships, as determined in independent UPS studies, are shown[20] in Fig. 8.4. The discrepancies between the band edge off-sets as determined by Parker[27] are small (~0.1 eV), and the band off-sets from the Fermi energy in the metals are in the same directions.

The transition from the UPS model to the LED model, which occurs during the PVD of the counter (i.e., Ca, in the present case)

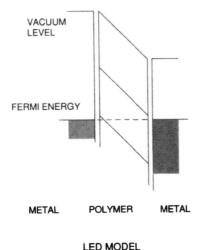

LED MODEL

Fig. 8.3   *LED band edge model, under conditions where the electro-chemical potential is a constant throughout the sample.*

electrode, is illustrated in Fig. 8.4. Provided that there are no surface states induced in/on the molecular solid (by rough treatment, or otherwise) at the base (Au) electrode[17,28], this effect appears to correspond to the charging of a capacitor[29], as illustrated in Fig. 8.5. The charges on the electrodes are given by the difference in the work functions, φ, of the metals prior to contact (the initial difference in the Fermi energies) and the static dielectric function of the polymer medium which fills the space in the capacitor, ε. From the relation $Q = CV$, where C is the capacitance (including ε) determined geometrically and $V = \Delta\varphi$ is the initial difference in the Fermi energies, the charge in the metal on the left (Au) is $-Q\varepsilon$, since electrons have transferred from the right-hand electrode with the lower work function (Ca); and the charge on the Ca-electrode is $+Q\varepsilon$. In the dielectric medium of the capacitor, the MEHPPV polymer, the charges are $+Q(\varepsilon-1)$ at the Au-electrode interface, and $-Q(\varepsilon-1)$ at the Ca-electrode[16,17].

The situation at the PVD electrode is not as simple as at the electrode upon which the polymer film was, for example, spin-coated. In chapter 7, it was shown how aluminum forms covalent bonds at the interface with PPVs and the model molecule DP7. Also, calcium dopes the near-surface region of clean (oxygen free) PPV, but forms an oxide on the surface of PPV with a significant amount of oxygen-containing species on the surface[30]. Again, without the loss of generality, the polymer–metal interface, at the electrode applied by PVD, may be illustrated as in the top of Fig. 8.6 (note the

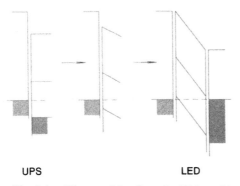

UPS   LED

Fig. 8.4 *The transition from the UPS model to the LED model.*

alignment relative to the vacuum level, so that the band edge energies are more obvious). At the top is the 'ideal' ITO–MEHPPV–Ca structure, used again as the representative structure. In the centre is the case where the Ca-metal has doped the near-surface region of the PPV, resulting in a thin region of 'conducting polymer', with filled bipolaron bands, in between the 'bulk' of the PPV film and the metallic portion of the Ca-electrode. From angle-dependent XPS, or XPS(θ), it can be seen that the Ca^{2+} ions are located near the surface of the PPV[31,32]. Although a good measure of the depth distribution of Ca^{2+} in PPV can not be given from these studies, the fact that a tendency to be localized near the surface in XPS(θ), with the

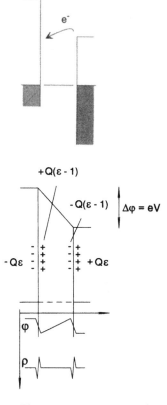

Fig. 8.5   *The charging of the LED capacitor. The charge density, ρ, corresponds to delta-functions, with the distance between the positive and negative peaks, illustrated at each interface, vanishingly small, i.e., on the order of an intermolecular distance.*

knowledge that XPS is only sensitive to the outermost 50–100 Å (as discussed above), indicates that the Ca^{2+}-ions reside more-or-less within a tunnelling distance of the surface, i.e., within say 20–30 Å, with some depth distribution. In the lower portion of the figure is illustrated the case of the oxide layer observed on surfaces of PPVs with a significant amount of oxygen-containing species[30]. Again the thickness of the interface layer is of the order of a tunnelling distance. In both of the cases illustrated, there is an interface region in between the pure calcium metal electrode and the bulk of the PPV film, which will affect the charge injection characteristics at these electrodes.

It is worth repeating the above. To date, the effects of the presence of interfacial barriers or interface regions have not been explicitly considered in models of charge injection at such interfaces. The

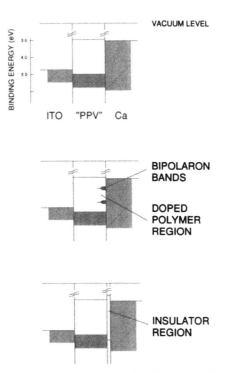

Fig. 8.6   *Ca–'PPV' interfacial barriers. The figures are drawn with alignment to the vacuum level of energy, so that the band edge energies can be seen. In this illustration, the electro-chemical potential is not a constant throughout the sample. The band edge values used for the polymer are for MEHPPV.*

presence of these interface regions will undoubtedly play an important role in ultimate device performance, and possibly even stability, in polymer-based LEDs in the future. The presence of the chemistry at these interfaces will certainly be the subject of intense study in the area of polymer-based LEDs in the coming few years, and may determine ultimate device behaviour.

8.5   Summary of typical conjugated polymer–metal interfaces

There appear to be two limiting cases to polymer–metal interfaces in the context of materials currently used in polymer-based light emitting diodes.

1 The metal is clean; the polymer is applied to the metal film, and there are no unsatisfied bonds on the polymer at the interface. In the absence of a counter electrode, the band edge off-sets are determined by the constancy of the electro-chemical potential, which lies in the middle of the HOMO–LUMO gap of the semiconducting polymer overlayer film (the UPS model). In the presence of a counter electrode, the band edge offsets are determined by the charging of a capacitor to a voltage equivalent to the difference in work functions of the two electrodes, and influenced by the dielectric medium (the conjugated polymer film) in between; a transition to the LED model; providing that there are no surface states at the substrate (first) electrode! There may be circumstances where interactions with the substrate do occur, however, even for spin-coated films. An example might be the adsorption of thiols or xanthate molecules[33] on gold surfaces or model molecules for polyimides adsorbed upon copper surfaces[34].

2 For metals which function as electrodes when vapor-deposited on to the surface of the conjugated polymer film, chemistry occurs. An intermediate layer is formed, between the electrode metal and the bulk of the conjugated polymer film. If the atoms of the metal electrode form covalent bonds, a layer of disrupted conjugation occurs, which in those cases investigated thus far, is of thickness on the order of a tunneling thickness. If the atoms of the metal deposited donate electronic charge to the polymer doping it to a

conducting state, then an interface layer consisting of a thin conducting polymer exists between the metallic electrode and the bulk of the conjugated polymer, the thickness of which also seems to be on the order of a tunnelling distance. For certain metal atoms, charge donation to the conjugated polymer occurs as the metal atoms diffuse uniformly (on very different time scales for different metal atoms) throughout the bulk of the polymer film, forming short circuits in LED applications. In many (but not all) cases, if there are oxygen-containing species on the polymer surface, either in the form of chemical side groups or as impurity molecular species, a metal oxide interfacial layer may be formed, the detailed nature of which depends upon the specific nature of the oxygen-containing species.

It is very important to note that we have never observed any indication that a simple metal–polymer contact can be formed by vapor-deposition without the formation of an interfacial layer, as detailed above.

8.6 References

1. R. Hoffman, *Solids and Surfaces: A Chemist's View of Bonding in Extended Structures* (VCH, New York, 1988).
2. M. Schott, *Private discussions* (1990).
3. W. R. Salaneck, N. Sato, R. Lazzaroni and M. Lögdlund, *Vacuum* **41**, 1648 (1990).
4. W. R. Salaneck, *Rep. Prog. Physics* **54**, 1215 (1991).
5. D. T. Clark and W. J. Feast (Eds), *Polymer Surfaces* (John Wiley & Sons, Chichester, 1978).
6. W. J. Feast and H. S. Monroe (Eds), *Polymer Surfaces and Interfaces* (Wiley, Chichester, 1987).
7. M. Schott, in *Organic Conductors: Fundamentals and Applications*, J. P. Farges (Ed.) (Marcel Dekker, New York, 1994), p. 539.
8. H. Lüth, *Surfaces and Interfaces of Solids* (Springer-Verlag, Berlin, 1993).
9. G. Gensterblum, K. Hevesi, B. Y. Han, L. M. Yu, J. J. Pireaux, P. A. Thiry, D. Bernaerts, S. Amelinckx, G. Van Tendeloo, G.

Bendele, T. Buslaps, R. L. Johnson, M. Foss, R. Feidenhans'l and G. Le Lay, *Phys. Rev. B* **50**, (16), 11981 (1994).

10. T. R. Ohno, Y. Chen, S. E. Harvey, G. H. Kroll, J. H. Weaver, R. E. Haufler and R. E. Smalley, *Phys. Rev.* **B44**, 13747 (1991).

11. A. J. Maxwell, P. A. Brühwiler, A. Nilsson and N. Mårtensson, *Phys. Rev. B* **49**, 10717 (1994).

12. P. Rudolf and G. Gensterblum, *Phys. Rev. B* **50**, 12215 (1994).

13. H. B. Callen, *Thermodynamics* (John Wiley & Sons, New York, 1960).

14. M. Sachs, *Solid State Theory* (McGraw-Hill, New York, 1963).

15. C. Kittel, *Introduction to Solid State Physics* (John Wiley & Sons, New York, 1986).

16. S. M. Sze, *Physics of Semiconductor Devices* (Wiley-Interscience, New York, 1981).

17. E. H. Rhoderick and R. H. Williams, *Metal-Semiconductor Contacts* (Oxford University Press, Oxford, 1988).

18. P. Nielsen, D. J. Sandman and A. J. Epstein, *Solid State Commun.* **17**, 1067 (1975).

19. C. Ziegler, *Private communication* (1995).

20. N. Johansson, *Private communication* (1995).

21. D. Braun and A. J. Heeger, *Appl. Phys. Lett.* **58**, 1982 (1991).

22. J. H. Burroughes, D. D. C. Bradley, A. R. Brown, R. N. Marks, K. Mackay, R. H. Friend, P. L. Burn and A. B. Holmes, *Nature* **347**, 539 (1990).

23. A. Holmes, D. D. C. Bradley, A. R. Brown, P. L. Burns, J. H. Burroughes, R. H. Friend, N. C. Greenham, R. W. Gymer, D. A. Halliday, R. W. Jackson, A. Kraft, J. H. F. Martens, K. Pichler and D. W. Samuel, *Synth. Met.* **55-57**, 4031 (1993).

24. C. W. Tang and S. A. VanSlyke, *Appl. Phys. Lett.* **51**, 913 (1987).

25. C. Adachi, S. Tokito, T. Tsutsui and S. Saito, *Jpn. J. Appl. Phys.* **27**, L269 (1988).

26. A. D. Brown, D. D. C. Bradley, J. H. Burroughes, R. H. Friend, N. C. Greenham, P. L. Burns, A. B. Holmes and A. Kraft, *Appl. Phys. Lett.* **61**, 2793 (1992).

27. I. D. Parker, *J. Appl. Phys.* **75**, 1656 (1993).

28. L. J. Brillson, in *Handbook on Semiconductors,* P. T. Landsberg (Ed.) (North-Holland, Amsterdam, 1992), p. 281.

29. G. Hansson, *Private communication* (1995).
30. Y. Gao, K. T. Park and B. R. Hsieh, *J. Chem. Phys.* **97**, 6991 (1992).
31. P. Dannetun, M. Lögdlund, R. Lazzaroni, C. Fauquet, C. Fredriksson, S. Stafström, C. W. Spangler, J. L. Brédas and W. R. Salaneck, *J. Chem. Phys.* **100**, 6765 (1994).
32. P. Dannetun, M. Fahlman, C. Fauquet, K. Kaerijama, Y. Sonoda, R. Lazzaroni, J. L. Brédas and W. R. Salaneck, in *Organic Materials for Electronics: Conjugated Polymer Interfaces with Metals and Semiconductors*, J. L. Brédas, W. R. Salaneck and G. Wegner (Eds) (North Holland, Amsterdam, 1994), p. 113.
33. A. Ihs, K. Uvdal and B. Liedberg, *Langmuir* **9**, 733 (1993).
34. W. R. Salaneck, S. Stafström, J.-L.Brédas, S. Andersson, P. Bodö and J. Ritsko, *J. Vac. Sci. Tech.* **A6**, 3134 (1988).

Index

Printed in the United States
By Bookmasters